Transforming Science and Engineering

Transforming Science and Engineering

ADVANCING ACADEMIC WOMEN

Abigail J. Stewart, Janet E. Malley,
& Danielle LaVaque-Manty

EDITORS

THE UNIVERSITY OF MICHIGAN PRESS   ANN ARBOR

Copyright © by the University of Michigan 2007
All rights reserved
Published in the United States of America by
The University of Michigan Press
Manufactured in the United States of America
⊗ Printed on acid-free paper

2010   2009   2008   4   3   2

A CIP catalog record for this book is available from the British Library.

Library of Congress Cataloging-in-Publication Data

Transforming science and engineering : advancing academic women /
    Abigail J. Stewart, Janet E. Malley, and Danielle LaVaque-Manty,
    editors.
        p.   cm.
    Includes index.
    ISBN-13: 978-0-472-11603-4 (cloth : alk. paper)
    ISBN-10: 0-472-11603-7 (cloth : alk. paper)
    1. Women in science—United States—Social conditions.   2. Women
scientists—United States—Social conditions.   3. Women in
engineering—United States—Social conditions.   4. Women engineers—
United States—Social conditions.   5. Women in higher education—
United States—Social conditions.   6. Women college teachers—United
States—Social conditions.   7. College science teachers—United
States—Social conditions.   I. Stewart, Abigail J.   II. Malley, Janet
E., 1950–   III. LaVaque-Manty, Danielle, 1968–

Q130.T73   2007
305.43'5—dc22   2007003220

This book is dedicated to the memory of Denice Denton,
whose advocacy and leadership on behalf of women
in science and engineering made such a difference.

Contents

Foreword

AS A SCIENTIST who obtained a Ph.D. in biochemistry in 1969, I have
long been keenly aware of the stark disparity between the number of
women who pursue science and engineering in graduate school and yet
who fail to emerge as faculty in our nation's colleges and universities,
particularly those with strong research missions.

In fact, in my experience, which I know is not unique, many women
have historically chosen to study science, but have not enjoyed an environment that encouraged pursuit of an academic career. In college, I
was one of ten chemistry majors, fully half of which were women.
However, no women were on any of the science faculties of the college,
a point that was not lost to any of us. In graduate school, a similar situation prevailed. No women were part of the faculty, but almost half of
the graduate students in biochemistry were female. At each stage, from
B.A. to Ph.D. to postdoctoral positions, more and more women left science or elected to take positions outside the academic realm. Certainly
in the 1960s there were very few role models for women students even
at a time when women were interested in preparing themselves for
careers in science.

Fortunately, an event occurred in my life that completely changed
the dynamic for my personal calculations about a career. Late in my
graduate career, Dr. Mary Ellen Jones, a very distinguished biochemist,
joined the faculty of the University of North Carolina as a consequence
of her husband's acceptance of a position of chair of the Department of
Pharmacology. Her presence in the Biochemistry Department was elec-
trifying for all the women students. Mary Ellen understood the power of
role models and so she accepted the role of chair of the committee to

invite external seminar speakers for the department. For the next two years, virtually all of the scientists who visited to give talks were prominent women. In a departure from past practice, she also made sure that all the graduate students were invited to speak with visitors on topics ranging from the science they were presenting to how they managed complex and busy lives. Those experiences were enormously influential for me as I know they were for many of my colleagues.

This was a vivid demonstration of the fact that if students never observe successful women faculty members, it is hard to envision themselves as a member of any faculty body.

Throughout my own career as a scientist, a faculty member, and a university administrator, I have been involved with and initiated many studies about the leaky pipeline for women academics. Again and again, I have confirmed through these studies cultural and institutional impediments to full participation by women. The resistance to change is profound and occurs at many, many different levels, a number of which I have experienced in my own faculty career. The nature of the entrenched culture in universities is why the interest and engagement of the National Science Foundation (NSF) in the ADVANCE Institutional Transformation program was so compelling. Alice Hogan, the leader of ADVANCE within NSF, had the vision and the understanding that a sustained and comprehensive national program was essential to produce information that could be scaled at institutions across the entire country.

Furthermore, I believe that the imprimatur of the NSF was critical to making a study of systemic change regarding participation by women within educational institutions a legitimate research exercise. Attractive financial resources were provided for successful applications and a robust mechanism to share data and information was an integral part of the entire project plan. After five years of research at a variety of institutions, data about the underlying causes of under-representation of women and effective strategies for systemic changes are finally available.

What Abby Stewart, Janet Malley, and Danielle LaVaque-Manty have accomplished so well within this volume is to present a roadmap for institutions to emulate regardless of where they are positioned on the spectrum of representation of women in science and engineering. They posit two overarching issues that are not often recognized. The question of what it means for an institution to have too few women isn't often addressed. Even rarer is for anyone to ask what effect that has on the women who are already on the faculty. The second is that features of the institutional environment will be important in whether women faculty thrive in scientific disciplines once they are on the faculty. Understanding and addressing both of these issues is critical in developing intervention strategies that work and that can be transported across institutions. Particularly appealing to me, as a university president, is the inclusion of a section within this book that surveys many ADVANCE institutional strategies and identifies those that are effective and attainable for a reasonable investment.

I found the research and development of intervention strategies contained in this volume compelling and directly targeted at the interest of our nation in securing adequate numbers of scientists and engineers for our future well-being. Study after study has warned of the growing person power crisis for our country. Tapping the talents and interests of women as well as men into the STEM disciplines—science, technology, engineering, and medicine—is part of the solution for our continued competitiveness in a global economy based on innovation and technology. I urge my presidential colleagues to read this book, evaluate their own institutional commitment to transforming science and engineering in order to attract and advance women academics, and implement effective strategies that have emerged as a consequence of the investment the NSF has made in the ADVANCE program. Our nation will be better for our willingness to halt the brain drain of highly accomplished women.

Mary Sue Coleman
President
University of Michigan

 PART ONE

Context

ONE

Analyzing the Problem of Women in Science and Engineering

WHY DO WE NEED INSTITUTIONAL TRANSFORMATION?

Abigail J. Stewart, Janet E. Malley, & Danielle LaVaque-Manty

ACADEMIC SCIENCE and engineering suffer from a peculiar sort of brain drain in the United States: they persistently lose highly accomplished women, not to other countries, but to industry and unrelated careers. Increases in the proportion of women faculty in tenured or tenure-track positions at research universities lag far behind the increases in the proportion of science and engineering doctorates that have been awarded to women in recent decades. The problem is clearly not that women cannot (or do not want to) conduct scientific research at the highest levels. What, then, is it?

When the National Science Foundation (NSF) announced the ADVANCE Institutional Transformation program, an NSF program designed to cultivate the success of women in academic science and engineering, it stated that "women scientists and engineers continue to be significantly underrepresented in some science and engineering fields and proportionately under-advanced in science and engineering in the Nation's colleges and universities. There is increasing recognition that the lack of women's full participation at the senior level of academe is often a systemic consequence of academic culture. To catalyze change that will transform academic environments in ways that enhance the participation and advancement of women in science and engineering, NSF seeks proposals for institutional transformation" (National Science Foundation 2001). This statement expressed concisely a perspective that shifted attention away from remedying real or imagined deficits in women (e.g., lack of suitable ability, personality, ambition, or willingness to work) as the cause of women's relatively low participation in science and engineering, and toward problems at the institutional level. This was not the first time a call for "institutional transformation" with respect to colleges and universities had been issued (see, e.g., Bleier 1984; Hubbard 1990; Schiebinger 1989), but it was a very significant one, in that it focused on the faculty, on science and engineering, and came from a federal funding agency. Though the problematic climate for women (students and faculty) in science and engineering had been discussed for decades (see, e.g., Hall and Sandler 1982, 1984), NSF's new program both legitimated the importance of that discussion as an explanation for the continued low representation of women scientists and engineers on college and university faculties, and called for action. Moreover, it backed that call with funding resources. In short, NSF not only changed the conversation about this issue, but also provided significant incentives for institutions to make serious efforts to change.

In order to make a serious effort to change, though, institutions needed to develop analyses of the problem. The funded institutions represented in this volume (and no doubt many that were not funded, or were funded in later rounds of the program) drew on a wide range of theories of organizational change, and human behavior, as well as many empirical studies, in developing their own analyses. They began searching for ideas in isolated social scientific studies of science (see, e.g., Sonnert and Holton 1995; Zuckerman, Cole, and Bruner 1991), as well as in studies of related issues in nonscience settings (Kanter 1977). Many quickly found that the best place to begin to develop an integrated understanding of the important factors minimizing women faculty's participation in science, or those that might maximize it, was Virginia Valian's volume *Why So Slow?* (1999). Valian brought together a wide range of social science studies of cognitive processes affecting evaluation that might account for women's very slow progress in virtually all professions in the second half of the twentieth century. Her book provided outstanding conceptual tools for many ADVANCE Institutional Transformation project efforts to understand women's experiences in science and engineering.

Perhaps the most central concept in articulating the need for "institutional transformation," at least in the domain of women in science and engineering, is the notion of *underrepresentation* itself (see Kuck 2001; Nelson 2005). This concept can be defined in different ways. For example, it could be argued that women should participate in every activity in the society in rough proportion to their numbers in the population (thus, about half). Alternatively, it could be argued that women should be expected to participate on university faculties in rough proportion to their attainment of doctoral level degrees (allowing for changes in the rates of that attainment). In the case of women faculty in science and engineering, women are "underrepresented" either way. But because university faculty members in all fields must hold higher degrees, ADVANCE projects have mostly operated from the latter definition. Like NSF itself, they have noted that in virtually all science and engineering fields, there are relatively few tenured women faculty at research institutions (Nelson 2005). But does this constitute underrepresentation? In deciding whether it does, projects also must consider a range of issues: what is the proportion of women students at every level in the present institution? What are the consequences for the students of having the faculty look very different from the student body? What was the proportion of women students getting doctoral degrees in various disciplines both at this institution and nationally five, ten, and twenty years ago? Does the faculty represent those figures or are women underrepresented on the faculty compared with past degree attainment too? (See Hopkins 2006 for a thoughtful discussion of these issues.)

There is no one right way to assess the adequacy of women's representation on science and engineering faculties, but the process of reviewing the data is an important element in coming to grips with the nature of the problem. When an institution can satisfy itself that the proportion of women on the faculty does mirror national levels of degree attainment (in any particular field or even across the institution, as for example, at Hunter College; see Rabinowitz and Valian, this volume), then the issues facing it are quite different from those facing an institution that draws a different conclusion. If there are too few women on the faculty, then hiring women becomes an important priority—and in many instances so does working harder to recruit more women into the field at earlier educational stages.

What does it actually mean for an institution to have "too few women"? We have noted (like Hopkins 2006) that it may suggest a distortion in past hiring, and it can have implications for students. But how does it affect the women who are actually on the faculty in those fields? At least two concepts have proven extremely important in thinking about this issue: *token* or *solo status;* and lack of *critical mass*. The literature on "solos" or individuals who are the sole representatives of their group (by race, gender, or other defining characteristic) suggests that they are

perceived and treated differently than individuals who share a characteristic with other members of their group in a work setting (Niemann and
Dovidio 1998; Yoder and Sinnett 1985). Overall, solos—particularly
solos who are members of lower-status groups in the society—are subject to more stereotyping, scrutiny, and negative judgments (Thompson
and Sekaquaptewa 2002). Not surprisingly, they also experience greater
internal stress. The notion of solo status has been extended, in an
attempt to understand the point at which group membership stops being
noticed—when "critical mass" is attained (see, e.g., Etzkowitz et al.
1994). Although there is no definite cutoff point, literature suggests that
when a group becomes a significant "minority" (perhaps as large as a
third of a larger group), individuals are viewed through a more individ-
ualistic (less stereotyping) lens. Thus, belonging to a particular socially
defined group (e.g., by gender or ethnicity) matters less under these
conditions—both to perceivers and to the person being perceived (see
Valian 1999, 139ff., for a review).

The second overarching concept underlying institutional transformation efforts is the notion that *features of the institutional environment* itself
matter; this concept links the failure of women to thrive in science,
technology, engineering, and mathematics (STEM) fields not to their
own inadequacies but to features of the workplace in which they operate (Johnsrud and Des Jarlais 1994; Long 2001; Preston 2004; Wright et
al. 2003). Particular concepts that have been used to understand and
study institutions include the *climate, tacit knowledge* (of how the institution, discipline, funding agencies work), *networks* (and exclusionary
processes), *schemas* (associated both with gender and with the fields of
study), *evaluation bias,* and *accumulation of disadvantage.*

Most of the ADVANCE institutions recognize that whether or not
women are underrepresented in science and engineering, the *climate* for
their effort is often "chilly" or not welcoming. That is, women are often
treated—by students, colleagues, or staff—as in some way unsuitable for
the work, and therefore untrustworthy as authorities, undesirable as colleagues, and not fitting in (Ambrose et al. 1997). This perception, when
communicated to women, can create and feed self-doubt and alienation.
Even when this perception is not directly communicated, it often leads
to women being excluded from networks by which important information about how institutions work, as well as opportunities, are communicated (Hitchcock et al. 1995). Equally, being viewed through the lens
of gender results in men's accomplishments in science and engineering
being routinely overvalued and women's undervalued (Banaji and

Hardin 1996; Banaji, Hardin, and Rothman 1993; Steinpreis, Anders, and Ritzke 1999). Annoying but possibly minor in a single instance, this pattern of evaluation bias—multiplied on many different occasions of evaluation of lectures, manuscripts, grant proposals, and so on—results in the *accumulation of significant advantage* to men and disadvantage to women over a lifetime. Valian has identified this as the process of creating mountains (of disadvantage) out of molehills (1999, 4–5). Institutionally the "chilly" workplace environment is communicated through policies and practices that assume that faculty members are not primary caregivers for others, and that they can rely on others to take care of those responsibilities in their lives (Preston 2004; Xie and Shaumann 2003). The gendered career trajectories produced by institutional biases result in a particular dearth of women in leadership roles in STEM fields, which of course symbolically legitimates the gendered system itself, and helps it to reproduce itself.

One important advantage to recognizing the institutional basis of women's underrepresentation and difficulty thriving in science and engineering (that is, of employing *systems theory* in at least a loose way to understand the problem; see von Bertalanffy 1975; Weick 1976; Weick and Orton 1990) is that it shifts the burden of guilt off of individuals. Just as women cannot be blamed for their own exclusion from science and engineering, men (and the few women on the "inside") cannot be blamed for it either, as individuals. If the problem reproduces itself unless it is interrupted (an inevitable feature of a system's self-maintenance once set in motion), then change can be initiated by individuals, but the problem is not individuals' fault. It is worth noting that the system operates at many levels (particular institutions, higher education as an institution, disciplines themselves, funding agencies, etc.). Given that fact, it is important that actions be considered that would address these issues at multiple levels. Most of the interventions discussed in this volume aim at particular institutions; one (Croson and McGoldrick's chapter) aims at a discipline (economics), but it is possible to imagine interventions (including, for example, the use of Title IX advocated by Senator Wyden; Mervis 2002) that would aim at other, broader levels, and that might be equally or more important. Indeed, the ADVANCE program itself is a national-level "intervention."

Many of the institutional transformation efforts described in this volume are based on analyses that rely on the concepts outlined above. Four key elements of their efforts are directly aimed at many of the obstacles already outlined. These include (1) identifying *norms or practices* in science and engineering generally, in a particular field, or in their institution that tend to exclude women; (2) *educating* individual faculty (women and/or men) or *raising their awareness* of the nature of the problems facing women in STEM fields; (3) *altering the representation of women in STEM fields* by increasing their numbers; (4) taking deliberate actions to *counter gender schemas and evaluation bias.*

In creating interventions that have these goals, projects rely on a range of different kinds of approaches or strategies. For example, many programs—implicitly recognizing women scientists' and engineers' isolation from one another, small numbers in many locations, and exposure to unwelcoming environments—aim to *create and support the development of women scientists and engineers' positive collective identity.* These efforts draw on what is known about the creation of groups or movements that can advocate on their own behalf and provide positive support for one another (Apfelbaum 1979).

Other interventions make direct *connections between individuals' personal experiences and policies and practices* in the institution. Thus, if women faculty are much more likely than men to carry a "second shift" of caregiving responsibilities, then institutional policies and resources need to be directed to recognizing that fact and supporting women faculty as well as their male colleagues with different household demands (Rapoport et al. 2002; Williams 2000).

Because men are colleagues and bosses of the women scientists and engineers, it is important to include men—both as leaders and as allies— in the process of institutional transformation. Many of the institutional change efforts rely on this recognition and attempt to foster and support *change-focused alliances of women and men* (Bolman and Deal 1991; Smith 1989; Chesler 1981). It is very important that these alliances be able to identify actions they can take to improve the situation; one critical aspect of intervention is *creating meaningful avenues for concrete action,* for alliances, but also for individuals and for the institution. When individuals' awareness of a problem is increased, but they are not helped to identify things they can do to address the problem, they can simply feel paralyzed by their new knowledge.

In the context of many of the interventions that have been identified, three broad principles are frequently noted to be helpful. First, many faculty and administrators assume that practices that enhance diversity will compromise excellence. It is critical to address this assumption, and to show how any given intervention designed to enhance diversity is compatible with high standards and the maintenance of excellent quality. Second, many of the actions taken to improve the environment for women have *benefits for most or all of the faculty*. If this is not explicitly stated, faculty are prone to viewing benefits in zero-sum terms, assuming that if things get better for women faculty, they will inevitably get worse for men. Third, progress (or its absence) cannot be assessed without *systematic monitoring of indicators*. It is very important to collect and use data to identify where progress is more and less substantial. These data need to be sensitive to the different levels at which change is wanted and may be occurring: in the women scientists themselves (their representation and the climate for them; their collective identity, raised consciousness, perceived linkage of personal experience to policy); in their male colleagues (in their raised awareness, concrete actions, and new alliances); in departmental and institutional leaders (in their awareness and actions); in institutional policies (for example, in their linkage of the personal to policy; in efforts to counter evaluation bias, and to assess and monitor indicators); and in organizational change leaders or "organizational catalysts" (who are identifying norms and practices that need adjustment and identifying avenues for action; see Sturm, this volume).

This volume, then, is intended as an inevitably partial compendium of institutional transformation efforts in the area of women faculty in science and engineering. Because it includes projects from a wide range of types and sizes of institutions, we believe it will be useful to individuals in an equally wide range of types and sizes of institutions. Of course, all of the institutions represented here both proposed and received "institutional transformation" grants from NSF, and that means they may differ from other institutions in several ways. First, they had committed organizational change agents on campus; second, they had substantial enough institutional support for the effort to make a credible proposal to NSF; and finally, they were provided (by NSF) with significant resources that allowed them to experiment with institutional transformation.

We believe that even without those resources other institutions can benefit from what these institutions have tried. There are, though, some minimum conditions that will make others more able to make successful use of these experiments. First, it is crucial to conduct a local analysis of the nature of the problem at the particular institution. Not only must remedies be designed to ameliorate particular problems, but one remedy will not suit all settings; all remedies require adjustment to local conditions. Second, the institution must have—in some key locations—

faculty and administrative leaders who have a sincere desire to make changes. Finally, there must be some energy and leadership that can be devoted to the transformation effort; it cannot happen without resources of time and leadership. There are, of course, other helpful adjuncts. It is extremely helpful if some financial resources can be made available to the project, and can support staff and others' time. It is also extremely helpful if the leadership of the institution at all levels understands and supports the institutional change effort. But we believe that meaningful change can happen—at least at some levels of the institution—even absent those adjunct conditions.

The book is divided into four parts. The first includes not only this introduction but an interview with the visionary leader of ADVANCE at NSF, Alice Hogan, who explicates the historical and institutional context for this program at a national level. The second part includes accounts of programs that provided enhanced or new institutional supports to women scientists and engineers themselves. These programs "warm up" the climate for women and help them thrive. The third part includes accounts of interventions aimed at transforming practices at the heart of academic institutions—hiring, evaluation, and tenure review. Finally, the fourth part provides analyses of the conditions for successful change. It includes case studies of particular circumstances leading to change (one entire institution and one department), of organizational catalysts and their role in institutional change, of the circumstances associated with institutionalization of change, and an account of the value of assessing different kinds of outcomes. The last chapter addresses a deeply practical question: it surveys the preceding efforts and identifies the ones that are readily adaptable, effective, and attainable at relatively low cost.

Providing Institutional Support to Women Scientists and Engineers

This second part of the book focuses on efforts to provide necessary, but previously missing, institutional supports to women scientists and engineers. These strategies differ from more conventional efforts to "fix the women" (rather than change the institution) because they are grounded in an analysis of an *institutional* failure to provide necessary conditions for women faculty members' successful growth and development. While each of the approaches offered in this part was specifically designed and tailored for the particular institution, because they address problems that arise for women scientists and engineers in many academic institutions, we believe they have real promise for adaptation in other institutions.

Many ADVANCE Institutional Transformation projects have tried to address one implication of women scientists and engineers' underrepresentation—their isolation from one another, and experience of being "tokens"—by creating stronger communities, or networks, among them. The specific approach offered by Rankin, Nielsen, and Stanley in chapter 3 was developed at the University of Colorado, and is built on the notion that new networks are tools for social change (just as traditional networks maintain the status quo), in part because they can increase members' understanding and capacity to challenge the ordinarily unchallenged assumptions about "how things work" in the institution, as well as how science works more broadly. From their perspective, then, it is important that networking not only shifts women faculty members' experience of themselves as members of a minority and stigmatized group, but more importantly that networking generates both new knowledge and increased ability to critique unquestioned norms, practices, and policies.

Dyer and Montelone, from Kansas State University, provide an account of a networking program with very specific goals for women faculty in STEM fields. Their aim was not to address women's lack of connection to one another on campus, but rather their lack of strong professional networks outside the institution. This program focuses on untenured women faculty—at the most vulnerable and least connected period in their careers—but clearly it could also operate for more senior women. This program provides individual women faculty with resources for creating a new and strong link with distinguished senior scientists from other campuses. In this case the goal is to provide individual women with the opportunity to develop a stronger professional network that in turn will provide increased mentoring, professional visibility, and external opportunities.

"Interconnected networks" are viewed as a critical element in Georgia Tech's institutional change effort. Realff, Colatrella, and Fox argue that it is useful to support a range of different sorts of networks, operating at different levels of the institution. As in the previous chapters, networking is expected to help address some of the consequences of skewed gender ratios in science and engineering fields, because "it allows those who would be excluded to overcome exclusion, connects strivers to role models, and fosters mentoring." In addition, though, they indicate that the networks fostered by their ADVANCE Institutional Transformation activities have also facilitated changes in faculty evaluation procedures, and the prevalence of reliance on gender schemas, while catalyzing collective action by women faculty.

While the networking programs outlined in these three chapters all have implications for career advising, Posey, Reimers, and Andronicos in chapter 6 describe a program at the University of Texas at El Paso designed much more focally to improve the mentoring received by women STEM faculty. Building on a recognition of the importance of preventing "the accumulation of disadvantage" in women's careers, the program experimented with different structures for mentoring, including two (mentors) on one (mentee) and mentoring pairs. Their recommended model is a small-group mentoring model, involving up to six mentees with two senior mentors, with assignments across related fields. As a result, though this program's explicit goal is mentoring, the structure increases networking too.

Rabinowitz and Valian outline a much more extensive "sponsorship" program initiated at Hunter College. The target group for this program is STEM women faculty of all ranks, who are offered structured support for their career development. Individuals are provided with workshops that offer precisely the sort of tacit knowledge about institutional norms and disciplinary practices that women may normally not have; in addition, resources are provided to ensure expert individual career advice from outside the institution; and a process of goal-setting and goal-monitoring is created. Women are encouraged to think about their careers in the context of their whole lives, and to address obstacles to their career achievement that arise in their personal lives, as well as to maintain a focus on balancing their work and family lives. This is an intensive program that has the capacity to have a major impact on a relatively small number of individual women who are, in turn, expected to become agents of change in the institution more broadly.

A very different kind of program to support women faculty was initiated at the University of Washington. The focus of the program outlined by Riskin, Lange, Quinn, Yen, and Brainard is women faculty members' occasional but critical need for support through difficult life transitions. The program recognizes both that women normatively engage in more caregiving than men, and that academic career tracks have little "give" to accommodate the disruption imposed by intensive parenting of a newborn, or caregiving for seriously ill family members. Specific resources provided are negotiated based on the faculty members' needs, as is the size of the award (from $5,000 to $38,000). The chapter provides information about related programs on other campuses, suggesting that this kind of support is widely needed and valued.

Transforming Institutional Processes

Chapters in part 3 report on programs designed to change key academic processes and academic climates. Rather than providing direct support to individual women faculty, these efforts target disciplines, institutional practices, or administrators, but they generally take place concurrently with efforts (like those described in the previous part) to enhance the careers of individual women, with the understanding that women's careers, like men's, will flourish only in institutions with equitable policies and practices and hospitable climates.

Stewart, Malley, and LaVaque-Manty, from the University of Michigan, outline an intervention designed to foster equitable approaches to hiring and recruitment through the creation of STRIDE (Strategies and Tactics for Recruiting to Improve Diversity and Excellence), a highly trained faculty committee that teaches administrators and recruitment committees about gender schemas, evaluation bias, and the accumulation of disadvantage. When the scientists and engineers who participate in STRIDE first came together to develop a new approach to faculty recruitment, they engaged in a sustained period of self-education, a period some of them now describe as "consciousness raising." Because they see themselves as people whose consciousness needed to be raised, STRIDE members are well positioned to explain gender schemas and evaluation bias to others without imputing any sense of blame for these phenomena. Their credibility (as trusted and institutionally supported scientists and scholars who transmit reliable theory and evidence) is crucial to the program's substantial success in developing change-focused alliances among men and women while providing departments with tools that will help them hire enough women faculty to reach critical mass.

In "Scaling the Wall: Helping Female Faculty in Economics Achieve Tenure," Croson and McGoldrick, from the University of Pennsylvania and the University of Richmond, describe an intervention that addresses an important question: what can be done to help women gain access to mentoring and tacit knowledge when they are a small minority within their departments or disciplines? Like Dyer and Montelone, whose chapter appears in part 2, they respond by helping women faculty develop ties that extend beyond a single institution, but in this case they build a discipline-wide network by holding mentoring workshops at national conferences. They also develop an intriguing concept they call "role activation"—the idea that those who have been mentored learn the value of mentoring and become more willing to adopt the role of "mentor" later when invited to do so. This is a program with the potential to create and sustain clear paths for the transmission of tacit knowledge among women faculty who might otherwise remain isolated as "solos" within their home departments.

Fox, Colatrella, McDowell, and Realff, from Georgia Tech, report on a method of addressing evaluation bias throughout their institution in "Equity in Tenure and Promotion: An Integrated Institutional Approach." Their program draws attention to the importance of many facets of credibility, from "leadership and organizational climate that signal the importance of equity," to the collection and presentation of relevant institutional data. Their project began with a comprehensive canvass of tenure and promotion policies across all units and a faculty survey "on issues pertaining to evaluation, including: resource allocation and success; mentoring and networking; perceptions of evaluative methods and procedures; interdisciplinary collaboration; entrepreneurship; and institutional culture." Based on their findings from this research, they developed a web-based instrument called "ADEPT" (Awareness of Decisions in Evaluating Promotion and Tenure) that not only conveys information about evaluation bias to those evaluating others for tenure and promotion, but also provides them with an opportunity to practice evaluating sample candidates and receive feedback on their reasoning so they will be better prepared to avoid evaluation bias when deciding the fates of real faculty. In addition, ADEPT transmits what would otherwise remain tacit knowledge about tenure and promotion processes to those undergoing evaluation. The ADEPT instrument could be of use to many other institutions and is available online: http://www.adept .gatech.edu/download.htm.

In "Executive Coaching: An Effective Strategy for Faculty Development," Bilimoria, Hopkins, O'Neil, and Perry, from Case Western Reserve University, outline an approach to improving the academic climate for women faculty by providing targeted training—coaching—to actors at many institutional levels. Women faculty are coached toward developing successful academic careers, while deans and department chairs are guided toward promoting a workplace culture "characterized by equality, participation, openness, and accountability," to create a climate that is hospitable to all faculty, particularly women. Coaches help deans and chairs understand obstacles for women faculty and attempt to "shift the underlying assumptions of the university culture." Though these institutional actors are coached one-by-one, the hope is that their changing administrative approaches will have a cumulative effect on the climate university-wide.

In "Interactive Theater: Raising Issues about the Climate with Science Faculty," LaVaque-Manty, Steiger, and Stewart describe an intervention designed to generate conversations about evaluation bias and departmental climate among faculty and administrators (without using those terms directly) through the use of sketches that present common departmental dynamics involved in gender and hiring, mentoring, and promotion. Though the sketches depend on live performance and facilitation of audience discussions, they share some key characteristics with Georgia Tech's web-based ADEPT tool, in that both are designed to offer decision makers the opportunity to practice thinking about evaluation bias with respect to hypothetical candidates, and to discuss their reasoning with others and receive feedback. The hope in each case is that those who have been exposed to the concept of evaluation bias and have a chance to practice applying what they've learned will be better equipped to avoid or counter bias when making real decisions later. This program has the potential to reduce the chilliness of departmental climates and reduce faculty reliance on gender schemas when evaluating women faculty.

Learning from Change

The typical ADVANCE project involves concurrent interventions taking place at multiple institutional levels. In contrast to chapters in the previous parts that select certain strategies or interventions from a larger set for close analysis, those in this part take a broader look at systemic interactions and conditions that foster transformation across entire departments or universities. They consider, too, means of institutionalizing change and ways of assessing outcomes across a range of different institutions that use incongruent categories, data, and measures. The final chapter in this part surveys a range of interventions tried by many ADVANCE schools and identifies several that might be easily adapted and that generate substantial change at relatively low cost.

Beginning with a focus at the departmental level, Jordan and Bilimoria of Case Western Reserve have conducted a case study of a science department that has succeeded in developing a workplace that is not only collegial and hospitable for women, but also highly academically productive. Believing with Meyerson and Fletcher (2003) that "gender inequity is rooted in our cultural patterns and therefore in our organizational systems," their chapter, "Creating a Productive and Inclusive Academic Work Environment," sets out to answer this guiding question: "How is a cooperative, inclusive, productive work culture created and embedded?" They find that most of the faculty in the department they studied have "belief orientations that supported more gender-integrated conceptions of who a scientist is and what a scientist does." In other words, they seem unusually free of gender schemas that associate science with men and masculinity. Members of this department also go out of their way to make sure that tacit knowledge is conveyed not only to junior faculty, but also to students and postdocs. Intriguingly, the entire department appears to have a culture of collaboration rather than competition, to promote a collective identity, and even to make having good social skills one of their criteria when hiring new colleagues. This chapter raises the possibility that cultivating better social practices can improve not only gender equity and workplace climate, but also the quality of the scientific research conducted within an academic unit.

Moving to a larger institutional perspective, in "Advancing Women Science Faculty in a Small Hispanic Undergraduate Institution," Ramos and Benítez ask how women faculty can advance "in an institution where advancing seems difficult for all faculty (including men)?" Constraints that contribute to difficulties for all faculty (but especially women) at the University of Puerto Rico–Humacao (UPRH) include disadvantageous hiring practices in which faculty positions are not publicly advertised, heavier teaching loads than those usually found in larger universities on the U.S. mainland, a lack of financial resources with which to buy equipment, geographic and linguistic isolation that makes it difficult to attract potential faculty from outside Puerto Rico, and a culture that strongly associates women with traditional family roles. This chapter describes a set of interventions designed to address all of these difficulties simultaneously, by offering grants and release time for individual women faculty, employing a legal advisor who educates faculty and administrators about gender issues, cultivating networks of women faculty and research collaborations between women faculty and their male colleagues, and developing an internal pipeline of future women faculty through a program called Faculty in Training, or FIT, in which undergraduate women serve as research assistants to women faculty at

UPRH and are helped through the graduate school application process, with the hope that they might be hired as UPRH faculty members after they complete their doctorates.

In "Gender Equity as Institutional Transformation: The Pivotal Role of 'Organizational Catalysts'," Susan Sturm argues that "the building blocks of systemic change are present in many institutions" and that "systemic change occurs through connecting the knowledge and action of . . . engaged participants." Based on her findings from a case study conducted at the University of Michigan, Sturm believes that NSF ADVANCE programs have put key actors whom she calls *organizational catalysts*—principal investigators on ADVANCE projects, among others—in positions that allow them to intersect multiple systems and bring influence and resources together to enable change. In addition to mobilizing varied forms of knowledge—ranging from data-gathering methods, to social science theories about gender, to formal and informal institutional policies and practices—organizational catalysts are able to develop collaborations in strategic locations and create and maintain pressure and support for action. Her chapter concludes with the suggestion that organizational catalysts could become a significant element of other change initiatives, and an examination of the risks and promises of attempting to institutionalize the organizational catalyst's role, a possibility Sturm regards as being "essentially an institutional design problem" that might, with care, be overcome.

In "Institutionalization, Sustainability, and Repeatability of ADVANCE for Institutional Transformation," Rosser and Chameau from Georgia Tech provide a set of questions designed to help readers considering organizational change projects think about their institutions in systemic and strategic terms, to determine whether conditions that would foster and support change are present at their universities. The compatibility of a project's goals with institutional values and leadership and the nature of the resources that would be available to support the change effort must be considered in order to decide whether the project is viable in the long term. "Honest, serious answers to these questions," they believe, "will help institutional leaders and potential principal investigators determine their readiness to develop a project in terms of scope, timeliness, and financial practice changes that ultimately will be sustained and institutionalized."

Frehill, Jeser-Cannavale, and Malley investigate the progress the first round of nine ADVANCE awardees made in hiring women faculty during their first few years of NSF funding in "Measuring Outcomes:

Intermediate Indicators of Institutional Transformation." They first discuss not only how to compare these institutions to one another across differences in size and geography, but also how to assess their progress against national data. The chapter offers an insightful review of ways to determine relevant levels of analysis when trying to understand whether women are in positions similar to men in various fields, and how to choose appropriate comparison groups when attempting to discern whether an institution is doing better or worse than its peers with respect to hiring and retaining women. Their analysis is a useful example of how to begin assessing the impact of attempts at institutional transformation while those attempts are still ongoing.

Lee Harle's chapter, "Maximizing Impact: Low-Cost Transformations," written while Harle was an AAAS/NSF Science and Engineering Policy Fellow at the National Science Foundation, should be particularly useful to those who wish to improve their institutions on small budgets. Sections of Harle's chapter address programs directed to faculty; programs designed to help administrators increase their understanding of gender and science so they might improve the climate for women faculty at their universities; and programs that can be implemented institution-wide. The chapter offers brief descriptions of each type of program, and in many cases, directions for how to find more information or gain access to web publications and other resources that various ADVANCE institutions have made available.

The interview with NSF program officer Alice Hogan that follows this introduction presents some of NSF's hopes and intentions for the programs described in this book. As Hogan notes there, readers "may be able to use this book very effectively by discussing these chapters and determining which elements they can use directly and which they might want to modify . . . The book allows all of us to trace ADVANCE's learning curve."

REFERENCES

Apfelbaum, E. 1979. Relations of domination and movements for liberation: An analysis of power between groups. In *The social psychology of intergroup relations,* ed. W.G. Austin and S. Worchel, 188–204. Belmont, CA: Wadsworth.

Ambrose, S. A., K. L. Dunkle, B. B. Lazarus, I. Nair, and D. A. Harkus. 1997. *Journeys of women in science and engineering: No universal constants.* Philadelphia: Temple University Press.

Banaji, M., and C. Hardin. 1996. Automatic stereotyping. *Psychological Science* 7 (3): 136–41.

Banaji, M., C. Hardin, and A. J. Rothman. 1993. Implicit stereotyping in person judgment. *Journal of Personality and Social Psychology* 65 (2): 272–81.

Bleier, R. 1984. *Science and gender: A critique of biology and its theories on women.* New York: Pergamon.

Bolman, L. G., and T. E. Deal. 1991. *Reframing organizations: Artistry, choice and leadership.* San Francisco: Jossey-Bass.

Chesler, Mark. 1981. The creation and maintenance of interracial coalitions. In *The impact of racism on white Americans,* ed. B. Bowser and J. Hunt, 217–44. Beverly Hills: Sage.

Etzkowitz, H., C. Kemelgor, M. Neuschatz, B. Uzzi, and J. Alonzo. 1994. The paradox of critical mass for women in science. *Science* 266:51–54.

Hall, R. M., and B. R. Sandler. 1982. *The campus climate: A chilly one for women?* Washington, DC: Association of American Colleges.

———. 1984. *Out of the classroom: A chilly campus climate for women?* Washington, DC: Association of American Colleges.

Hitchcock, M. A., C. J. Bland, F. P. Hekelman, and M.G. Blumenthal. 1995. Professional networks: The influence of colleagues on the academic success of faculty. *Academic Medicine* 70 (12): 1108–16.

Hopkins, N. 2006. Diversification of a university faculty: Observations on hiring women faculty in the schools of Science and Engineering at MIT. *MIT Faculty Newsletter* 18 (4): 15–23.

Hubbard, R. 1990. *The politics of women's biology.* New Brunswick, NJ: Rutgers University Press.

Johnsrud, L. K., and C. D. Des Jarlais. 1994. Barriers to tenure for women and minorities. *Review of Higher Education* 17 (4): 335–42.

Kanter, R. M. 1977. *Men and women of the corporation.* New York: Basic Books.

Kuck, V. 2001. Refuting the leaky pipeline hypothesis. *Chemical and Engineering News* 79 (47): 71–73.

Long, J. S., ed. 2001. *From scarcity to visibility: Gender differences in the careers of doctoral scientists and engineers.* Washington, DC: National Academy Press.

Mervis, J. 2002. Can equality in sports be repeated in the lab? *Science* 298:356.

Meyerson, Debra E., and Joyce K. Fletcher. 2003. A modest manifesto for shattering the glass ceiling. In *Reader in gender, work, and organization,* ed. R. J. Ely, E. G. Foldy, M. A. Scully, and The Center for Gender in Organizations, Simmons School of Management. Malden, MA: Blackwell Publishing.

National Science Foundation. 2001. ADVANCE: Increasing the participation and advancement of women in academic science and engineering careers. Program announcement nsf0169.http://www.nsf.gov/pubs/2001/nsf0169/nsf0169.htm.

Nelson, D. J. 2005. A national analysis of diversity in science and engineering faculties at research universities. http://cheminfo.chem.ou.edu/ ~djn/diversity/briefings/Diversity%20Report%20Final.pdf.

Niemann, Y. F., and J. F. Dovidio. 1998. Relationship of solo status, academic rank and perceived distinctiveness to job satisfaction of racial/ethnic minorities. *Journal of Applied Psychology* 83:55–71.

Preston, A. 2004. *Leaving science: Occupational exit from science careers.* New York: Russell Sage Foundation.

Rapoport, R. A., L. Bailyn, J. K. Fletcher, and B. H. Pruitt. 2002. *Beyond work-fam-*

ily balance: Advancing gender equity and workplace performance. San Francisco: Jossey-Bass.

Schiebinger, L. 1989. *The mind has no sex? Women in the origins of modern science.* Cambridge: Harvard University Press.

Smith, D. G. 1989. *The challenge of diversity: Involvement or alienation in the academy.* Washington, DC: George Washington University.

Sonnert, G., and G. Holton. 1995. *Who succeeds in science? The gender dimension.* New Brunswick, NJ: Rutgers University Press.

Steinpreis, R. E., K. A. Anders, and D. Ritzke. 1999. The impact of gender on the review of curricula vitae of job applicants and tenure candidates: A national empirical study. *Sex Roles* 41 (7–8): 509–28.

Thompson, M., and D. Sekaquaptewa. 2002. When being different is detrimental: Solo status and the performance of women and racial minorities. *Analyses of Social Issues and Public Policy* 2:183–203.

Valian, V. 1999. *Why so slow? The advancement of women.* Cambridge: MIT Press.

von Bertalanffy, L. 1975. *Perspectives on general systems theory.* New York: George Braziller.

Weick, K. 1976. Educational organizations as loosely coupled systems. *Administrative Services Quarterly* 21:1–19.

Weick, K., and J. D. Orton. 1990. Loosely coupled systems: A reconceptualization. *Academy of Management Review* 16 (2): 203–23.

Williams, J. 2000. *Unbending gender: Why family and work conflict and what to do about it.* New York: Oxford University Press.

Wright, A. L., L. A. Schwindt, T. L. Bassford, V. F. Reyna, C. M. Shisslak, P. A. St. Germain, and K. L. Reed. 2003. Gender differences in academic advancement: Patterns, causes, and potential solutions in one U.S. college of medicine. *Academic Medicine* 78:500–508.

Xie, Y., and K. A. Shauman. 2003. *Women in science: Career processes and outcomes.* Cambridge: Harvard University Press.

Yoder, J. and L. M. Sinnett. 1985. Is it all in the numbers? A case study of tokenism. *Psychology of Women Quarterly* 9:413–18.

Zuckerman, H., J. T. Cole, and J. T. Bruner, eds. 1991. *The outer circle: Women in the scientific community.* New Haven: Yale University Press.

TWO

Transforming the Scientific Enterprise

AN INTERVIEW WITH ALICE HOGAN

Danielle LaVaque-Manty

ALICE HOGAN has served since 1999 as chair of the ADVANCE Coordination Committee (the "design" committee) and the ADVANCE Implementation Committee, and became the program officer for ADVANCE in 2001. This interview was conducted by telephone in September 2005.

Danielle LaVaque-Manty: Could you talk about NSF's rationale for creating ADVANCE? What made NSF commit to it, or you commit to it? What was the thinking about what needed to happen?

Alice Hogan: NSF had long been interested in increasing the participation of women and other groups in academic science and engineering, but a convergence of several trends led to the creation of the program now known as ADVANCE. NSF had for over a decade run programs for women in recognition of the fact that women were not yet fully part of the science and engineering enterprise—the premise was that women didn't have access to the connections, the knowledge, or the inclusion in informal networks it takes to launch and sustain successful research careers. Earlier NSF programs sought to provide women access to research funding that would allow them to emerge as independent researchers. In fact (somewhat contrary to expectations and to community belief), the proposals submitted under the auspices of these "women's" programs were generally so competitive that many proposals were ultimately funded through the general funding pool because of their high quality.

In addition to this unexpected pattern of very high quality proposals and relatively low "success rate," there were several problems with this set-aside approach. The pot of available money was relatively small, making it unlikely that real change could be effected through these programs. Funding enough research proposals for enough individual women to make a real difference would have required a huge amount of money. In addition, over the years, there were a number of different versions of programs for women. NSF had a tendency of starting these programs and then stopping them rather abruptly to try something else. So they would gear the community up and then say, "Wait, we're going to try something different." There was an informal perception within NSF and perhaps within the larger community that this happened more frequently with programs for women than with other NSF programs.

What NSF eventually realized was that it wasn't that women in particular needed help accessing research funds, but that the system they were working in was littered with obstacles to their success. Individual research grants could only reach a limited number of people, and while these grants were often powerfully effective for those individuals, the environment those people worked in was not changing and often did not support their success.

DL–M: Did something prompt that? Did something happen that caught NSF's attention, or was this strictly an internal, organic shift?

AH: NSF had been thinking about increasing participation in science as a systemic issue for some time, but unlike the issues for other groups underrepresented in science and engineering, where people were not choosing science in college, women *were* well represented in the pipeline. Women were choosing science as undergraduates, pursuing graduate work in science, and earning doctorates in science. The part of the system that needed attention was the academic workplace.

In July 1999, a group of program directors at NSF was invited to recommend strategies to NSF management that would be the basis for a program to foster the advancement of women in science and engineering faculty ranks, especially at senior leadership levels.

The development of ADVANCE involved discussion among all the disciplinary areas funded by NSF, literature reviews, and an assessment of what was being done systemically to address the factors that limited women's full participation in science and engineering. The committee found that the MIT report on the status of women, released in May

1999, provided a valuable perspective on what might be done and the level of effort involved in working at the institutional level.

DL-M: It sounds like NSF sees science as a system, or a culture, or something beyond a series of knowledge-building fields. We might have expected NSF to think in terms of geology developing "geology knowledge," chemistry developing "chemistry knowledge," and so on. But it sounds like NSF has a perspective on science as a human system that is more than just a knowledge factory. Is that right?

AH: NSF may be unique among the federal agencies that deal with science, because its original mandate not only gave it the responsibility to fund research, but also tasked it explicitly with ensuring the national supply of scientists and engineers. So NSF necessarily focuses on integrating research and education, along with the human dimensions and the broader impact of science. The decision to develop a program that would be called ADVANCE was both consistent with NSF's mandate and an evolution of the foundation's thinking about how to broaden participation in science.

DL-M: Could you say something about the funding for ADVANCE? Where does the money come from?

AH: To ensure that the work was adequately supported, we wanted to put enough money behind the call to make generating those proposals worth doing. The foundation was committed to providing adequate support to allow people to take the time to think creatively about changing institutional practices and policies, and to providing enough funding to support the time of the people engaged in this work. The budget allotted to ADVANCE in the first year was $9 million, and in subsequent years the budget increased to $19 million. These funds are collected from the research directorates at NSF and pooled for use by the ADVANCE program.

After a series of discussions with NSF senior management, the first call for proposals went out to the public early in 2001. Despite the fact that applicants had only three months to develop their proposals, a surprisingly high number of ADVANCE proposals for institutional transformation—seventy-two—were submitted.

DL-M: What were your expectations, or hopes, for the proposals?

AH: At NSF, we had no idea what to expect from these proposals. We knew we wanted to create an opportunity for people to think creatively, to take chances, and to create a community of people who took these issues seriously and had funding to tackle deep-rooted systemic barriers.

NSF seeks to work on the cutting edge of science and engineering, where innovation and risk go hand in hand. ADVANCE was risky, since there were so many unknowns. It was also innovative, in attempting to shift the focus to the systemic factors and away from the individual. While we hoped that writing the proposals might generate new ideas, we also believed that a major part of the innovation would be in implementing change in an institution-wide, high-profile way. Much was known about what the issues were, but we knew of few places that had attempted to tackle the issues this way at an institutional level. We believed that out of actually implementing real interventions would come greater knowledge about what worked, what didn't, and why. Our own deliberations were helped by looking at social science literature, and we hoped people might do that in their proposals as well, and that what they submitted would be based on a good understanding not only of faculty demographics, but also of the underlying causes of gender inequities in science. In the first round, the proposals that stood out were submitted by institutions that had already been thinking about these issues, that included social scientists as PIs [principal investigators] or as part of a development group, and that had already data collected or had an institutional profile of working on these issues.

The second round of competition followed fairly quickly on the first. By the second round, we were starting to ponder what would come next. We expected institutions to draw on what had already been done by those who had received ADVANCE grants in the first round, but we also wondered if it made sense to keep funding more and more institutions to try slightly different things. There was no way that NSF could fund each institution that needed to address the institutional factors limiting women—there are far too many.

DL-M: Did you find that the proposals offered new ideas? Did this elicit new strategies, or was it more that it focused institutional commitment?

AH: The question that persists is, what is new? What is still out there that we don't know? Are most of the interventions that seem to be succeeding simply common sense, good practice in terms of management and communication? Are these things that all universities should obviously be doing in order to be better workplaces for everyone? Or is there something new to learn about the gendered nature of these processes and how they work? We also recognized that we still were not seeing proposals that had fully developed strategies for supporting the participation and advancement of women of color in academic science and engineering, which left significant numbers of people out of the ADVANCE community.

The announcement for the third round of ADVANCE grants went out earlier this year, with much greater specificity in the call for proposals, both to clarify expectations and to reduce the number of proposals that ignored the work already underway in ADVANCE. The level of sophistication in the third round of proposals is quite high. The idea that there is already a body of knowledge out there is well accepted now. People who might not have known this in the past are now giving us extensive citations of existing work, and framing the conceptual design of their proposals with reference to current scholarship and practice.

DL-M: Part of what I'm hearing from you is that the desire to identify new understandings and new strategies still drives ADVANCE, but that there is also a desire to identify effective or successful strategies from among those that already exist. Is that right?

AH: There are interesting challenges for ADVANCE as the program progresses. One is getting the word out so that the resources developed through program investments are adapted and used widely. That is where we have high hopes for the volume in which you plan to publish this interview. We don't want people to keep reinventing the same wheel, but on the other hand, we wonder how much of the knowledge that will be reflected in the book will have to be rediscovered by an institution on its own to really be understood. There may need to be a sense of ownership through discovery. Even so, we hope and anticipate that groups may be able to use this book very effectively by discussing these chapters and determining which elements they can use directly and which they might want to modify. The book could be a huge time saver for institutions that want to change for the better through interventions of their own, but don't want to plow through all the primary literature that's out there.

In reflecting on the evolution of the ADVANCE program, it is amazing and inspiring to see what a transformative effect NSF has had with respect to this issue by simply saying that the system of science has a gender problem. The fact that the foundation sought ideas for transformation, and committed substantial resources, has allowed people who would never have dared (or dreamed, depending on who they are) to discuss the issues with their colleagues. Women who may have been suffering but who knew they would activate all the stereotypes if they said one word about it, now can bring this up as a funding opportunity. Women faculty who had been isolated, as the first and often only woman in their departments, also found through ADVANCE a network of other women faculty in science and engineering at their institution, allowing them to see that what they may have thought was their problem alone was in fact a systemic problem affecting many. Men, department chairs, for example, who might have wondered why all the women on their short lists turned down their job offers, also have a chance to see how other departments have become departments that attract the top women candidates, and learn about resources that will show the way to greater success in recruiting and retaining faculty.

NSF has been gratified to learn that the process of applying itself can make a difference. The proposal preparation process requires getting buy-in across a number of departments and colleges, getting commitments from upper administration to sign on, and meeting with colleagues whom you might never have met before. Simply submitting a proposal can help an institution begin to develop its own useful network and to become aware of other networks that are already out there. This effect helps us at NSF feel slightly better about the fact that many deserving proposals go unfunded in this era of flat budgets.

It is clear that some strategies have been working better than others, and this book carries important information about what strategies have been working best in certain settings and under various conditions. The book allows all of us to trace ADVANCE's learning curve.

DL-M: What has been the most striking thing you've observed as you've watched the ADVANCE programs at various universities establish themselves?

AH: What is exciting about where we are now—and I wouldn't want to say that it is unexpected, but we didn't anticipate it happening as powerfully or as quickly as it has—is the development of the network of people who are thinking critically about advancing women faculty. The network includes people both on individual campuses and across the nation who would never have known each other if they hadn't begun trying to improve things for women in science. It's an amazingly extensive and cross-disciplinary network, in fact, it is best described as a network of networks. This impact of ADVANCE may be one of the most enduring and fascinating changes brought about through ADVANCE and the innovative and courageous work of the ADVANCE institutions. This book itself is a demonstration of the impact of those kinds of networks. In order to be at the point of being able to put out, in a chapter or article, an informed analysis of specific elements of interventions or strategies that have worked particularly well, these authors have had to draw on all kinds of capital at their own institutions to get buy-in to try these things, bringing in faculty that they maybe never would have interacted with otherwise and administrators they probably didn't even recognize before they began.

The social capital that is reflected in the development of this book, along with the growth of understanding that has gotten us this far, is enormous. And this is only one product of what we at NSF expect to be a continually growing and learning community, because there is a whole community of institutions and individuals that people can call upon now that didn't exist before. Those who want to be part of this network, those who are doing this work and drawing on information that has already been put out informally, on web sites, for example, can look to this book for a deeper understanding than they may have been able to find previously.

The impact of the work ADVANCE has fostered, of which this book is one manifestation, has been to create a remarkable community of people who might never have expected to be part of something like this—people from all kinds of disciplines and career stages, both male and female. This book represents the maturity of an unexpectedly fruitful collaboration.

PART TWO

*Providing Institutional Support to
Women Scientists and Engineers*

THREE

Weak Links, Hot Networks, and Tacit Knowledge

WHY ADVANCING WOMEN REQUIRES NETWORKING

Patricia Rankin, Joyce Nielsen, & Dawn M. Stanley

WEAK LINKS and hot networks are key parts of a strategy for institutional transformation that has emerged from the Leadership Education for Advancement and Promotion (LEAP)[1] project at the University of Colorado at Boulder. LEAP's goals are to produce the long-term institutional changes needed to improve faculty retention among women and minorities and to increase the representation of women and other minorities both in leadership positions and in the science, technology, engineering, and mathematics (STEM) fields.

We use the term *weak links* to describe interactions between individuals and groups that otherwise would not interact—individuals and groups that are different enough in background, training, and taken-for-granted assumptions that contact between them is likely to generate clashes between, and questioning of, paradigms, along with much exchange of information. This kind of interaction, then, if encouraged and fostered, has the potential for catalyzing major and minor change and as such is important to the goals of our program.

Hot networks are highly energized, remarkably productive teams that lead to unusually satisfying intense interactions (sometimes for an extended period). More to the point, faculty participants in hot networks are less likely to consider leaving for another university or another career track. Insofar as some agreement about basics—for example, the question or issue at hand—is necessary for hot networks to function, one might think that they are composed of like-minded people and thus quite distinct from networks formed by weak links. But hot networks are just as easily composed of individuals whose enthusiasm centers on new, recently adopted or modified ways of thinking. Thus in the dynamic world of network forming and transitioning, weak links and hot networks within a department, organization, or institution may occur independently or in tandem. The relative frequency of each and how they are related has to be determined empirically.

Tacit knowledge, a third major component of the project (and chapter title), constitutes a significant chunk (but certainly not all) of those academic and nonacademic paradigms that are currently operating in contemporary universities. We first relate the concept of tacit knowledge to networking and then focus on those aspects of tacit knowledge that we think are relevant to the success of women and minorities in STEM fields.

This chapter considers why networks should be considered key tools—both conceptually and in fact—for effecting the kind of cultural climate and social transitions required to reach the project's goals. Although we focus on networking and related processes in this chapter, there is a larger model of social change that we seek to develop. This model has both structural and process-based components. On the one hand, we recognize the importance of a system-wide approach such as that described by Austin (1998)—that is, many different, specific actions aimed at each and every level of organization in a larger system. At the same time, we note that the processes described here are more likely to be incrementally small changes that add up to synergistically larger impacts. They more closely resemble what Katzenstein (1990) describes as "unobtrusive mobilization," and what Meyerson (2003) calls "tempered radicalism." (See also Meyerson and Fletcher 2000.) Unobtrusive mobilization refers to conscious change efforts that are grounded in the daily experience of women located within institutions. It includes such strategies as negotiation, selective support, targeted undermining, diffusion of radical ideas, and raising gender consciousness, all of which Katzenstein calls the "politics of associationalism." The expression *tempered radicalism* captures a similar process of change but with an emphasis on women and men as leaders in business organizations. Our long-term theoretical goal is a synthesis of these potentially related and similar but not yet integrated themes (systemic change, unobtrusive mobilization, tempered radicalism) that are found in the institutional and organizational change literature as it relates to diversity. But first, a bit of history.

LEAP and Networking

The LEAP program began with a focus on development of the managerial and leadership skills of faculty members (both women and men) to better enable them to succeed in an academic setting. One tool for achieving this goal was to host two series of expert-led workshops specifically designed for junior and senior faculty, respectively. We considered that managerial and leadership training filled a gap in faculty development and that junior women in particular would benefit from being introduced to negotiation techniques, time management, and strategies for working effectively in the university setting. We had observed that women in departments that operated according to established "best practices" such as inclusion and transparency in decision making felt more welcome. We decided that encouraging the adoption of such practices by giving chairs the opportunity to learn them would improve the campus climate.

These workshops in turn generated two unanticipated developments. First was the informal but extensive formation of valuable, lively, and influential faculty networks by the faculty themselves. The second was that through discussion in these networks we became aware of the salience and widespread acceptance of assumptions about what faculty consider to be necessary for success in STEM fields. These networks are generally informal, and we became aware of them partly through the reports of our program evaluators and partly through comments made directly to us by program participants. LEAP often provides modest support for faculty get-togethers designed to discuss topics of mutual interest such as work life balance and child care options.

For example, many professors (especially new professors) and graduate students in STEM fields have internalized a set of assumptions about how to succeed professionally: work "24/7," do not have children, spend more time in the lab to do better-quality work, and accept that current demographics of the STEM workforce reflect a meritorious selection process. These assumptions—when taken seriously—have the effect of discouraging many women (and some men) from pursuing tenure-track careers in STEM fields. We'll say more about these assumptions and their role in this project. For now, some basics about networking.

Networks 101

A network, generically defined, is a web of relations through which people interact with one another (Scott 1991). Network studies categorize and document different types and ways that people connect. Networks can be social (bridge club) or professional (American Medical Association), and range from informal to formal, temporary to permanent, loose to tightly organized, and from multifaceted to specific in their goal orientation. In everyday life, of course, networking—whether within smaller units such as academic departments, in larger contexts such as institutions and bureaucracies, or across national and international boundaries—is dynamic and always in flux. Connections within networks, as well as those across networks, are relevant to this project.

Within a network, members can be described as being relatively central or peripheral to the network itself. Central people are those who have contact made with them or make contact with many others. They can be thought of as hubs of activity. Peripheral members are not strongly connected to the network and as a result are more isolated. Cross and Parker (2004) argue that peripheral members represent missed opportunities in business structures—that their expertise or knowledge is not being used. They also state that peripheral members are those most likely to leave an organization. Network centrality, then, is important for retention in academia. To the extent that women in STEM fields are more likely to have peripheral positions in academic networks, they are not only an underutilized resource but may be less than satisfied with their professional associations and thus more likely to consider leaving the institution, the discipline, or academia.

Connections between networks are described as ranging from "strong" to "weak." According to Lin (2001), strong ties link networks of people with similar resources who in many ways share an identity or group paradigm. Weak ties are looser connections between groups of individuals who may differ substantially in their backgrounds. Although weak ties rely less on strongly shared values, they often provide access to new types of information and resources. Granovetter (1973), for example, showed that short-lived weak ties were much better suited for finding job information than longer-term strong ties. This distinction has been further developed to emphasize the important role of weak ties in areas other than job searching and of people who single-handedly link one or more networks. The term *bridging* refers to networks joined by weak ties.[2] Our interest in bridging networks also reflects a relatively recent development in the literature that explores the role that networks and clusters play in promoting innovation and knowledge exchange. Freeman (1991), for example, in his description of "networks of innovators," emphasizes that external sources of scientific and technical information are crucial to a unit's successful innovation.

Networks can become "hot"—that is, its members develop an enthusiasm for some idea or project and become intensely excited about their participation in the group. Hot groups are, more often than not, spontaneous, fluid, small, and initially generated by taking on the solution of seemingly impossible problems. In this regard they foster innovative thinking. They are characteristic of, but not limited to, successful and highly productive collaborative research-oriented groups. Hot networks are described as rare, but we note that there are exceptions to this rule in certain environments.

In an early study of networks specific to academia, Price (1963) hypothesized the existence of "invisible colleges"—that is, an in-group of scientists that exists in each of the specialties of science. People in such groups claim to be reasonably in touch with everyone else who contributes materially to research in a particular field. They constitute a power group that controls personal prestige in the field and influences the fates of new scientific ideas. This claim has been modified by more recent researchers who argue that academic networks are not monolithic. Instead they tend to be composed of subunits, usually centered on one "high producer" who has contacts with other "high producers" (Crane 1972). Subunits often consist of senior researchers and junior researchers who were once their students. We speculate that many of these high-producer subunits are what we are now calling hot networks. How gender, race/ethnicity, and other status variables are patterned within these models of academic networking is less well developed but certainly worth exploring.

We hypothesize that weak links, which bring unlike minds together, and hot networks, which keep people intensely interacting, tend to be associated with the kind of changes and paradigm shifts that are relevant to large-scale change initiatives. These two phenomena (weak links and hot networks) can occur independently and stimulate change independently. They can also intersect, creating new and powerful clusters of activity. An intervention like LEAP, which encourages the development of new links through multidisciplinary workshops (along with leadership training), can help to create an environment in which bridging networks, which are more likely to challenge the status quo, can develop. The idea here is that when faculty participants are exposed to new information and less familiar paradigms, the likelihood of modifying, adjusting, changing, or at least questioning one's own familiar assumptions is increased. While change can occur in other ways in institutions, a strategy of creating bridging networks—whether hot or not—may be particularly relevant for LEAP program goals such as long-term institutionalization and broad-scale changes. Networks thus created have reverberating power insofar as their members take on leadership roles and serve on decision-making committees. They are then in a position to disseminate new information, generate thinking "out of the box," and function essentially as catalysts for change. As an example, one LEAP workshop participant worked with the University's diversity office and the Chancellor's Committee on Women to launch a series of focus groups designed to gauge the climate of the campus as it relates to women and minorities. Results of this effort were incorporated into the Chancellor's annual state-of-the-campus address and subsequent university policy. This can be considered a nice example of what Meyerson (2003) means by tempered radicalism. Tempered radicals effect institutional change by working within an institution as members of the institutional structure. These individuals can work effectively within a structure without necessarily accepting the status quo and current values of the organization.

Many of the LEAP-generated networks confer direct advantages on their members as sources of practical advice and basic information on how to get things done while working within the university bureaucracy. We hypothesize that these networks are especially important to women and faculty of color who might otherwise be more isolated and who may not be acutely aware of the importance of networking. Junior faculty in particular report that their peers are good sources of information and reassurance about the difficulties of being a junior faculty member. As one woman said, "It's good to know that I'm not the only person who finds this hard."

Networking also helps to spread information about university policies already in place. They help especially those protected-class faculty in departments whose chairs may not be particularly knowledgeable or who may be perceived as hostile to activities such as stopping a tenure clock. They provide useful alternative perspectives on academic life by giving people information about departments other than their own. As an example, we have discovered anecdotally that such networks make people aware of the possibility that a poor departmental climate can be changed. Likewise, library faculty are using LEAP-generated networks to inform others that they (library faculty) are subject to the same research requirements and standards for tenure as faculty in other colleges.

Factors Influencing the Formation of Networks: Minority Status, Gender, and Physical Location

Kanter (1977) showed that small groups of minorities (defined as being less than 15% of a workgroup) encounter greater discrimination (and therefore less access to collegial networks) through the mechanisms of isolation, boundary heightening (being subjected to negative comparisons with the majority such as "women don't do that kind of research"), and stereotyping. Kanter argued that as the percentage of the minority group increased, discrimination decreased, allowing greater access to networks. However, Yoder (1991) argues the opposite position: that as minority group size increases, the minority group experiences greater discrimination from a majority group that fears losing power. This suggests that change efforts need to be sensitive to the possibility of a backlash and should not expect that smooth transitions to broader representation will necessarily follow increases in numbers of women and minorities in a work unit. We speculate that both hypotheses are correct, depending on context: Kanter's model suggests assimilation on the part of minority members, while Yoder's model indicates minority members who maintain a strong gender or ethnic identity.

As might be expected, gender has been very carefully studied as a "minority factor" (Zuckerman, Cole, and Bruer 1991; Campbell 1988; Rose 1985; Reskin 1978; Kaufman 1978). Hitchcock et al. (1995) identify seven differences between the professional networks of men and women. Women have fewer male colleagues, fewer associates from previous institutions, and more friends (as opposed to colleagues) in their professional networks than do men. Unmarried women seem especially to be excluded from male networks and to have an even greater proportion of friends in their networks. Men are more likely than women to include superiors (faculty of higher rank) in their networks. Women also consistently rate their networks as less effective at helping them build a professional reputation than do men. Finally, women's networks are negatively affected when they have children younger than six and when they change jobs in response to their spouse's mobility, whereas men's networks are unaffected by these same events.[3] Whether similar generalizations can be made about underrepresented ethnic groups, sexual minorities, or disabled faculty has yet to be determined.

Research that focuses on the effect of space constraints on networking (Cross and Parker 2004) indicates that the ability to interact face-to-face, while not crucial for network formation, makes networking easier. Sharing the same physical space can encourage conversations (like chatting around the water cooler), and makes it more likely that a shared culture will develop (see Cross and Parker 2004 for the challenges faced by an international company in trying to create a "team" out of a group of people who were located in several countries). In the university setting, space constraints can affect networks. If a department is scattered across several buildings, it may be difficult for members to meet and interact regularly. Some members may feel that being placed in another building is a rebuke, or proof of "outsider" status. Also relevant to space and networking is the likelihood that those who are physically challenged face additional hardships in gaining access to buildings and finding appropriate office space.

When Networks Go Bad

As proponents of greater networking, we do not suggest that all networking is beneficial or that those individuals who have a central role in a network are always effective change agents. Cross and Parker (2004) demonstrate that more is not necessarily better, and that simply adding people to one's network does not necessarily improve either the quality of a network or the efficiency of a business. In addition, they declare that central people may represent roadblocks to conducting business if they have too great a gatekeeping role. For example, an administrator at a university is central if everyone has to receive permission for any and all curriculum changes, but the administrator could actually be slowing down innovative efforts on the part of others.

The work of Rose (1985) shows another downside to networking specific to women: as stated earlier, women are more likely to incorporate friends into their professional networks, and at the same time feel stymied in reaching professional goals. This finding suggests that it is important to network selectively and strategically with specific goals in mind. Not all networks are created equal, and some faculty may benefit from training on how to build effective professional networks.

The famed "old boy's network" is an example of a strong (versus weak) network. That is, the members share a common identity and common beliefs—beliefs that tend to discourage diversity in the workplace. Kanter's (1977) work demonstrates how these shared beliefs translate into actions such as stereotyping and isolating minority members. (One goal of the LEAP project, of course, is to undermine and disrupt these groups via extensive weak-tie networking.)

Networking also requires work. Hitchcock et al. (1995), summarizing other research, identify two additional "downsides" to networking: more negotiation (which can lead to greater expenses for phone calls and travel), and more time and emotional energy spent on maintaining good collaborative relationships among members.

Social Capital and Tacit Knowledge

Despite some negatives associated with networking, we assert that strategically increasing the professional networking done by women and other less well represented groups will have long-term beneficial effects. An extensive literature describes the benefits of networking for individuals as well as communities in terms of increased social capital (Field 2003). Social capital is defined by Putnam (1995, 67) as "features of social organization such as networks, norms, and social trust that facilitate coordination and cooperation for mutual benefit." Social capital refers to the resources an individual gains through networking with others; it is a measure of the value of knowing a lot of people, particularly people who can provide information, mentoring, and other contacts, all of which can contribute to the attainment of one's goals. Social capital is generated both by informal networks that develop over a period of time as people discover common interests, and through more formal structures such as an organizational hierarchy. The relation between social capital and bridging or bonding networks is less clear cut. Some writers (Bourdieu 1986) see social capital as belonging primarily to the privileged in society, such as "old boy network" members, and as a way of maintaining that privilege through strong (versus weak) ties. Others (Putnam 2000; Coleman 1988) point out that even distant ties can lead to an increase in social capital.

When studying social capital, researchers examine both the *process* and the *effects* of networking. That is, having access to networks and networking affects and increases one's social capital (see Field 2003); which means that the barriers women and minorities face in creating networks reduce their ability to accumulate social capital. We are particularly interested in a kind of social capital called *tacit knowledge.*

All bodies of knowledge consist in part of tacit rules that are difficult to articulate and thus left unverbalized. Within academic disciplines, tacit knowledge takes the form of information that is implicitly known by insiders and is so much part of the paradigm within which a field operates that the assumptions, procedures, and processes that constitute it are rarely stated and virtually never challenged. Collins (1982) gives as an example of tacit knowledge in math, the ability to distinguish between phrases in which x stands for an abstraction, the unknown in an equation to be solved, on the one hand, and phrases in which it is a specified, concrete value, on the other. Most of us switch back and forth between these two uses without being aware of a shift in thinking. Another example, again from mathematics, is place value in Arabic numbering. Place value is the value given to a number because of its position. The numerical phrase *12,* for example, is read as "twelve" rather than "one-two" or "one and two." Likewise, a digit placed to the left of two others has to designate hundreds. We read numbers this way without comment, but probably would notice our process more if we had to multiply and divide inside a different system—with Roman numerals, for example. (This example comes from Gombrich 2005.) Within the larger university structure, instances of tacit knowledge are hard to come by precisely because they usually remain unarticulated, but we offer as examples (1) knowing the difference between "soft" and "hard" money and (2) realizing the difference between power and authority in a bureaucracy and knowing to whom to go for an effective decision about, say, space. This kind of information is learned through social interaction, usually in the context of networks, and is used to navigate social structures, usually without a lot of conscious thought. Those who are outside a functioning network would have to ask for such information. Thus the extent of one's tacit knowledge varies considerably by social location as well as by time spent in an organization's culture. This leads us to a consideration of how tacit knowledge might be related to gender.

Tacit Knowledge and Gender

Tacit knowledge is relevant to the LEAP project in several ways. First, refer back to beliefs about the kind of life one needs to have to be successful in STEM fields, especially as a tenure-track scientist or engineer in academia. We have the impression—which needs empirical investigation—that women and men hold these assumptions differently, with men having a somewhat different context in which to hold them. The 24/7 mandate, for example, may be less of an issue to men than to women who are dealing with second-shift concerns. We now classify these beliefs as "unchallenged assumptions" and seek to catalyze discussion and deconstruction of these unstated and taken-for-granted assumptions about the nature of work in STEM fields. Closer examination of these beliefs and how they vary by gender, ethnicity, age, rank, and discipline will contribute to this process.

Another way tacit knowledge is relevant to the project has to do with its relative availability, especially to women and minority scientists. As an example, when one of the authors was looking for her first job as an assistant professor, she did not realize that the starting salary and startup package were negotiable. In fact, she did not learn so until later, when she discovered she was earning significantly less than more recently hired assistant professors. When she queried her postdoctoral advisor about why he had not told her to negotiate, he responded, "Everyone knows to do that." But she hadn't known, nor did many of the other women in academia she questioned about it. In fact, the literature shows significant gender differences in the use of negotiation (for a summary, see Babcock and Laschever 2003). The same author was reminded of her experience many years later when reading a dissertation on the differences between how men and women approached tenure (McClam 2004). One of the female postdocs McClam interviewed reported being frustrated because she was unable to duplicate a chemical synthesis described in a research paper, only to learn that a key step in the process had not been described precisely because it was assumed. In these two examples, neither the author as new assistant professor nor the frustrated chemist had access to key pieces of tacit knowledge directly relevant to success in their respective fields.

Becoming fully proficient in one's field involves accumulating tacit knowledge, and this may be especially important in the STEM fields, where the role of labs, demonstrations, and other kinds of hands-on knowledge is critical. One learns what it means to be a scientist and the cultural norms embedded in one's discipline along with learning how to do science. Gaining tacit knowledge sometimes requires face-to-face interactions—the ability to "watch/see/do" that is part of active involvement. At the very least, people need access to a behavioral model in order to emulate it. Much tacit knowledge is gained visually rather than verbally, which means that one-on-one or interpersonal networking is key to its acquisition. Connections between gender and access to tacit knowledge have not been extensively explored elsewhere but are an emergent theme in this project and deserve further inquiry. We hypothesize that one of the reasons it can be difficult to succeed in the STEM disciplines when following a nontraditional career path is that these less standard career paths make it less likely that one will have been able to acquire the discipline-specific tacit knowledge that is necessary to succeed.

Tacit knowledge is also relevant to interaction and success that depends on interaction outside of one's department or field—for one's career in the university setting. Here, weak links can be particularly important to women in STEM disciplines. Since female STEM faculty often find themselves the only woman in a department, or one of few, having connections with women outside the department can increase knowledge of policies that are more likely to affect women (e.g., stopping the tenure clock, sick leave, parental and family leave). Indeed, we have been surprised to learn how many women of childbearing age are not familiar with the university's family leave policy. We are pleased to report that one of the networks created by LEAP is a support group specifically designed for young faculty dealing with issues such as pregnancy and parental leave. As a result, sharing and disseminating relevant policy information has increased.

We close this section with a summary statement about the power of networks and tacit knowledge. As Maskel (2000) points out, there is value in not having to rely solely on formal mechanisms and procedures. Because much relevant knowledge is applied in nature, well-functioning networks provide their participants not only with the "know-what" but also its application in embedded settings ("know-how") by people who develop substantial but often tacit expertise ("know-who").

Networking for Change

For reasons elaborated above, increasing networking is now an explicit goal of LEAP. However, we have been surprised to discover that many women faculty are uncomfortable with increasing networking as one of the project's stated aims. At first glance, this is counter to the stereotype that women tend to be better at networking because of their social skills. We discovered, though, that many women believed that making connections with people in the workplace on the understanding that those connections could have a professional payoff in the future was manipulative and therefore not a fair or legitimate strategy for advancement. In other words, instrumental or goal-oriented networking seemed less acceptable than informal, psychosocial networking.

We addressed these concerns in several ways. First, we responded by writing a "how to" guide on networking (Rankin and Nielsen 2005), which gave practical advice on ways to broaden one's professional network as well as a commentary underscoring the importance of networking. Then we continued to develop our understanding of networks and their role in academia and research. Evidence suggests that the type and extent of networks within an institution play a key role in determining the success of change initiatives at that institution (Mohrman, Tenkasi, and Mohrman 2003). We are especially interested in the development of networks that prove to be major resources for faculty both by promoting research and scholarly creativity—e.g., hot networks—and by providing useful knowledge about the university as an organization.

Equally important is the use of empirical work generated by this project to contribute to theoretical development, specifically to the literature on social change via processes of everyday interaction. We are studying networks among junior and senior faculty, with a focus on how LEAP initiatives (such as workshops) can influence their growth and development. Indeed we have embarked on an extensive empirical study of networking with the goal of monitoring both the social processes involved and their effects. To that end we are tracking, recording, and observing workshop participants' networking activities and then mapping them in terms of organizational structures and hierarchies. The latter is guided by Austin's (1998) emphasis on targeting multiple levels of hierarchy within the institution, and connects to the overall strategy for change. Through questionnaires and interviews we are asking about network specifics—with whom one networks, whether they are on or off campus, whether they are within or outside one's own department, and so on. as well as about what networking means to the participants, how they view its importance in career success, and what helps or hinders their networking. A major task, of course, is to assess differences in networking by gender, discipline, and rank. This research will inform strategies for conveying the importance of networking, discovering ways to better incorporate women and minorities into networks, and using networks to both communicate tacit knowledge about university policies and challenge widely held tacit assumptions about how one succeeds in STEM fields.

How will this work complement that of others in this field? We said that the processes of unobtrusive mobilization and tempered radicalism described earlier were developed in the context of large-scale institutions (the military and the Catholic Church) and business organizations (corporations), respectively. Our work extends the study of similar change processes to an academic setting. By focusing on networking, which, though not excluded, is not featured in other studies cited, we extend the connection between networking and social change. We also extend this literature by specifying how tacit knowledge becomes an important form of social capital through networking. Not least in importance is gaining information about the role that networking can play in challenging and dismantling assumptions about science. And finally, the relation between gender and exposure to tacit knowledge has not, to our knowledge, been extensively studied.

Current Status

This is a work in progress, and we note some unresolved issues, tensions, and contradictions that we anticipate will be clarified over time with continued effort. First, we point to the fact that the project attempts to do both social action and basic research, both lofty goals but ones that often work in rather different practical directions. We are trying to change the world at the same time that we are trying to monitor it. Likewise, our reflexive use of praxis: theoretical work on networking has informed our strategies at the same time that results of our interventions have generated theoretical questions. This flexibility allows us to take advantage of potentially useful or profitable new directions but also makes for an unwieldy, unpredictable project.

Any multifaceted program aimed at institutional change faces the problem of pinpointing the incremental effects it has had, and LEAP is not unique here. It is difficult (if not impossible) to demonstrate conclusively that this project's actions, rather than a multitude of other factors, are the determinants of any advances or setbacks for women and faculty of color at our university. We also acknowledge that some possible project effects will be long term and may not be visible for years. Further, there are conceptual and definitional problems inherent in the literature on social capital and networking. To the extent that social capital and networking are defined in terms of each other and because both are both process and product, it is difficult to construct unambiguous and useful operational definitions of them. There are also problems associated with measuring such elusive concepts as tacit knowledge. This phenomenon is especially difficult because tacit knowledge is so rarely spelled out.

Thus we have our theoretical, conceptual, and methodological challenges. At the same time, resolving these issues can be a major contribution.

The 800-Pound Gorilla: Future Plans

The discussion thus far has focused primarily on two valuable aspects of the networks that LEAP has fostered: (1) network participants as disseminators of information that generates social action, and (2) LEAP graduates as unobtrusive or obtrusive, even organized, social change agents. We end with what we consider the key role these networks will ultimately play—as sites to launch examinations of popular beliefs about science. This will be the focus of the final years of the project in large part because the research we have been doing on why people make the career choices they do tells us that this is what must be changed to ensure the program achieves its long-term goals.

Debra Rolison (2004) has led discussions of the modern university's evolution from medieval monasteries without adjustment to the needs of the modern workforce. She points out that beliefs about how best to do science are so ingrained that no one challenges the idea that current practices are the most effective—that the evolution of the scientific method has stalled out. We modify this assertion by noting the postmodern and feminist literature that constitutes a critique of mainstream science (see Nielsen 1990) and add that a major purpose of this project is to revive this discussion and extend it to those in STEM disciplines. Moreover, with additional knowledge about networks specific to departments, we consider the introduction of weak links into existent STEM networks that will promote this discussion.

The existing paradigm of how STEM disciplines work is the 800-pound gorilla facing change initiatives. Until these tacit assumptions are examined, critiqued, and evolved, we will be offering palliatives rather than cures for the underrepresentation of women in STEM fields. Dramatic change requires a paradigm shift, and we are looking to networks to ultimately drive that shift.

NOTES

1. This program is funded by an ADVANCE Institutional Transformation award from the National Science Foundation, SBE 0123636. More information on the program can be found at http://advance.colorado.edu.

2. Other terminology includes "linking" or "loose" networks.

3. We note that although these differences are associated with gender, they are likely the result of women's roles and positions rather than anything inherent to being male or female.

REFERENCES

Austin, A. E. 1998. A systems approach to institutional change and transformation in higher education: Strategies and lessons from American and South African universities. Unpublished paper.

Babcock, L., and S. Laschever. 2003. *Women don't ask: Negotiation and the gender divide.* Princeton, NJ: Princeton University Press.

Bourdieu, P. 1986. The forms of capital. In *Handbook of theory and research for the sociology of education,* ed. J. G. Richardson, 241–58. New York: Greenwood Press.

Campbell, K. E. 1988. Gender differences in job networks. *Work and Occupations* 15:179–200.

Coleman, J. S. 1988. Social capital in the creation of human capital. *American Journal of Sociology* 94:95–120.

Collins, H. M. 1982. Tacit knowledge and scientific networks. In *Science in Context: Readings in the Sociology of Science,* ed. B. Barnes and D. Edge, 44–64. Cambridge: MIT Press.

Crane, D. 1972. *Invisible colleges: Diffusion of knowledge in scientific communities.* Chicago: University of Chicago Press.

Cross, R., and A. Parker. 2004. *The hidden power of social networks: Understanding how work really gets done in organizations.* Boston: Harvard Business School Press.

Field, John. 2003. *Social capital.* London: Routledge.

Freeman, C. 1991. Networks of innovators: A synthesis of research issues. *Research Policy* 20:499–514.

Gombrich, E. H. 2005. *A little history of the world.* Trans. C. Mustill. New Haven: Yale University Press.

Granovetter, M. 1973. The strength of weak ties. *American Journal of Sociology* 78:1360–80.

Hitchcock, M. A., C. J. Bland, F. P. Hekelman, and M. G. Blumenthal. 1995. Professional networks: The influence of colleagues on the academic success of faculty. *Academic Medicine* 70 (12): 1108–16.

Kanter, R. M. 1977. *Men and women of the corporation.* New York: Basic Books.

Katzenstein, M. F. 1990. Feminism within American institutions: Unobtrusive mobilization in the 1980s. *Signs* 16 (1): 27–54.

Kaufman, D. R. 1978. Associational ties in academe: Some male and female differences. *Sex Roles* 4:9–21.

Lin, N. 2001. *Social capital: A theory of social structure and action.* Cambridge: Cambridge University Press.

Maskell, P. 2000. Social capital, innovation, and competitiveness. In *Social capital: Critical perspectives,* ed. S. Barin, J. Field, and T. Schuller, 111–123. Oxford: Oxford University Press.

McClam, S. L. 2004. Fitting in or opting out: Deconstructing the marginalization of women in academic science. Ph.D diss., University of Colorado at Boulder.

Meyerson, D. E. 2003. *Tempered radicals: How everyday leaders inspire change at work.* Boston. Harvard Business School Press.

Meyerson, D. E., and J. K. Fletcher. 2000. A modest manifesto for shattering the glass ceiling. *Harvard Business Review* 7 (1): 127–36.

Mohrman, S. A., R. V. Tenkasi, and A. M. Mohrman Jr. 2003. The role of networks in fundamental organizational change: A grounded analysis. *Journal of Applied Behavioral Science* 39 (3): 301–23.

Nielsen, J. M, ed. 1990. *Feminist research methods: Exemplary readings in the social sciences.* Boulder, CO: Westview Press.

Price, D. J. 1963. Invisible colleges and the affluent scientific commuter. In *Little Science, Big Science.* New York: Columbia University Press.

Putnam, R. D. 1995. Bowling alone: America's declining social capital. *Journal of Democracy* 6 (1): 65–78.

———. 2000. *Bowling Alone: The collapse and revival of American community.* New York: Simon and Schuster.

Rankin, P., and J. M. Nielsen. 2005. Networking: Why you need to know people who know people. In *Success strategies for women in science: A portable mentor,* ed. P. Pritchard, 109–32. Amsterdam: Elsevier Science and Technology Books.

Reskin, B. 1978. Sex differentiation and the social organization of science. *Sociological Inquiry* 48:6–36.

Rolison, D. 2004. Can Title IX do for women in science and engineering what it has done for women in sports? http/www.case.edu/admin/aces/documents/Rolison%20Lectures.pdf. Accessed November 27, 2005.

Rose, S. M. 1985. Professional networks of junior faculty in psychology. *Psychology of Women Quarterly* 9: 533–547.

Scott, J. 1991. *Social network analysis: A handbook.* London: Sage.

Yoder, J. D. 1991. Rethinking tokenism: Looking beyond numbers. *Gender and Society* 5 (2): 178–92.

Zuckerman, H., J. Cole, and J. Bruer. 1991. *The outer circle: Women in the scientific community.* New York: W. W. Norton.

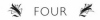

FOUR

An Institutional Approach to Establishing Professional Connections

Ruth A. Dyer & Beth A. Montelone

A NUMBER OF studies have emphasized the importance of professional networks in the success of faculty members (CAWMSET 2000; Pattatucci 1998; Sonnert 1995; Trower 2002). "The benefits of being in a strong network of contacts are the mirror image of the problems of isolation. Early inclusion in a strong network provides a 'jump-start' to a scientific career" (Etzkowitz, Kemelgor, and Uzzi 2000, 116). We have collected data that indicate that women faculty members in science, technology, engineering, and mathematics (STEM) departments at Kansas State University (K-State) lack access to such professional networks. We conducted an informal email survey of STEM women faculty members during the preparation of our proposal to the National Science Foundation ADVANCE Institutional Transformation program. One consistent element in their responses was the desire to reduce the isolation they felt in their departments and research disciplines. This isolation is the product of small numbers of women faculty in STEM disciplines, the relative remote geographic setting of our community, and the tendency of many departments to hire faculty members whose expertise does not overlap with that of others in the department. Male faculty members may experience feelings of isolation that are the result of the latter two items as well, but they generally have a wider pool of colleagues in their department with whom they share more common interests.

To address this concern, we considered possible ways to create linkages among women at K-State and their professional colleagues at other institutions. The most pressing need seemed to be that of untenured women faculty who had not yet established close relationships with colleagues in their disciplines. Noticing that senior faculty members often cultivated their connections with their colleagues via invitations to deliver presentations at departmental seminars, we decided to create a formal process by which these opportunities could be made available to the group with the greatest need. While it is true that tenured women may be more isolated than their tenured male colleagues, they generally are more likely to have professional networks in place because of previous attendance at national conferences and professional meetings.

Our ADVANCE Institutional Transformation project began in October 2003 and features a variety of project-wide, college-level, and department-level initiatives. One project-wide initiative is the ADVANCE Distinguished Lecture Series (ADLS), in which untenured women faculty members are given the opportunity to invite nationally recognized leaders (male or female) to our campus for an invited lecture. Eligible participants are tenure-track women faculty in all of the twenty-seven STEM departments in the four colleges that are part of the K-State ADVANCE Project. The purpose of this program is to assist junior faculty women in creating a network of colleagues in their professional disciplines who can provide guidance and mentoring in the establishment of their careers and foster their inclusion in wider professional circles. We believe such networks will lead to increased tenure, promotion, and retention rates of women faculty in STEM disciplines at K-State.

While these lectures are typically part of a standing lecture series for faculty and students within the department or college, the speakers also are designated as ADVANCE Distinguished Lecturers, and the K-State ADVANCE Project advertises the lectures throughout the university. During each visit to campus, the speaker may engage in a variety of activities such as visiting the host's research lab, working with the host to refine research techniques, or giving a guest classroom lecture. The host and the speaker may plan future meetings and collaborations. The speaker also interacts with other faculty members and students to discuss common research interests.

As part of the visit, the junior faculty member hosts the guest speaker at a dinner attended by her dean and department head. Funds for these dinners are provided by the college deans. The ADVANCE project provides partial support ($800 to $1,000) for honoraria and travel expenses for the speaker. The remainder of the support is provided by the host faculty member's department.

After the conclusion of the visit, each host faculty member submits a report to the ADVANCE project describing the activities in which the speaker participated during his or her visit, benefits accruing to the host as a result of the visit, and plans for future interactions with the speaker.

Local Context

As at many other institutions, women at K-State are underrepresented at all ranks in STEM departments and in administrative positions. In fall 2004, women constituted only 13.8% of the STEM faculty (tenured and tenure track) at K-State, with the majority of these women occupying the assistant and associate professor ranks; only 6% of full professors are women. A study of faculty cohorts in STEM departments hired in the years 1992–93 through 1997–98 showed that 61% of the women and 71% of the men achieved tenure (K-State ADVANCE Project Indicator Data 2005). The population of individuals who did not earn tenure includes those who left the university prior to the tenure decision as well as those who were denied tenure or reappointment. Longer-term retention rates, as of fall 2004, for STEM faculty are 51% for women and 58% for men for faculty cohorts hired between 1988–89 and 1997–98 (K-State ADVANCE Project Indicator Data 2005). This examination of historical tenure/promotion and retention data for men and women in STEM at K-State does not reveal statistically significant differences in tenure success or retention, but overall, these percentages are slightly lower for women than for men. The statistical nonsignificance may be influenced by the small number of women in these disciplines.

A dramatic increase in the percentage and demographics of women in STEM disciplines over the five-year funding period of our ADVANCE project would necessitate large increases in the numbers of women recruited and retained through tenure and promotion relative to men. This is unlikely to occur in such a short time period; however, we expect our ADVANCE initiatives to result in some short-term improvement in these numbers and have the potential for longer-term effects.

Implementing the New Strategy: Eligibility and Application Process

Eligibility. Tenure-track women assistant professors who are within five years of hire are eligible to participate in this program. Each faculty member can invite one speaker per year and can participate a maximum of five times or until tenure is granted. A call for proposals with guidelines for application is distributed as a hard copy and is posted online (K-State ADVANCE Project 2005) each fall and spring for seminars that would occur during the next six-month period. For each call for proposals we identify the set of eligible women faculty members, invite them all to apply, and ask the department heads to personally contact and encourage those who have not yet participated to apply.

Application and Review Process. The application process is deliberately simple. Each application must include the name and department of the woman applicant; the name, affiliation, and curriculum vitae of the proposed speaker; and a description of how the interaction with the speaker will potentially benefit the woman faculty member's career.

Remittance and Reporting Forms. The ADVANCE project web site also contains links to the reporting and remittances forms that are used to provide information about the visit to project staff and to request reimbursement for costs following the visit.

The K-State ADVANCE Steering Committee reviews the applications and requests additional information, if necessary. Applicants are notified of acceptance approximately two weeks after the deadline for submission of proposals. This rapid turnaround allows the women hosts to contact their speakers and arrange the visits in a timely manner.

Level of Participation

In the first two years of the ADLS program, eighteen women faculty members submitted applications to host a total of twenty-seven speakers. All of the applications were approved. Participation has included approximately 60% of eligible STEM women from about three-quarters of the STEM departments having eligible women faculty members. Of the eighteen women who have participated, five (28%) are women of color. Nine of the fourteen women who hosted a speaker in the first year also hosted a speaker in the second year, with two other hosts in the first year no longer eligible for participation because they were being considered for tenure. Table 1 summarizes the extent of participation in each of the first two years of the program.

Nineteen of the lectures took place in the first two years of the project. There were twelve men and seven women in this group of speakers. The majority of these speakers were from large research universities, including the University of California at Berkeley, the University of Georgia, the University of Maryland, and the University of Michigan. Other speakers came from a research hospital, a federal research laboratory, a consulting engineering firm, and a professional organization. Most of these individuals are full professors, some of whom hold named chairs or administrative appointments. All are nationally recognized leaders in their research specialties. A list of completed and upcoming lectures is maintained online (K-State ADVANCE Project 2005).

Reported Interactions with Speakers and Benefits to Host

Each faculty member who hosts a speaker provides a written report to the ADVANCE program that describes the types of interactions that occurred during the visit and planned future interactions. These written reports and verbal communications about the success of the visits from the hosts' department heads and deans have provided important feedback about the impact of this progam. These reports are, in general, highly positive and identify an extensive range of activities in which the speakers have participated.

We also obtained feedback through interviews conducted by the Office of Educational Innovation and Evaluation (OEIE), which is the evaluation unit that conducts assessment of our ADVANCE project activities. In these interviews, the host women were asked to describe the activities their speakers engaged in on campus, identify which were most and least beneficial to them, and share the reactions of their colleagues.

TABLE 1.   Level of Participation in ADLS

	Number of Eligible Women	Number of Departments Represented	Number of Applicants	Number of Departments Represented
Year One (2004)	23	15	14 (61%)	10 (67%)
Year Two (2005)	24	12	13[a] (54%)	10 (83%)

[a]Nine of these women hosted speakers in Year One.

Table 2 summarizes the activities in which the speakers participated while on our campus, as noted by the hosts in their self-reports and in the interviews conducted by OEIE as of October 1, 2005. Because of the timing of the lectures, this table includes data from eighteen hosts of the nineteen lectures that have occurred. The activities are listed in order from most frequently to least frequently mentioned by the host. Specifically, these include discussions of current research, suggestions for future research projects, review of the host faculty member's curriculum vitae and grant proposals, and demonstration of research techniques. Other hosts reported more informal interactions such as the speaker having a meal with students from the host faculty member's research program.

In describing which of these activities was the most beneficial, the respondents most commonly described the one-on-one interaction with the speaker. They also described the seminars and meetings between the speaker and the host's colleagues as particularly beneficial. The most well received activities included the speaker's seminar and the general interaction with the speaker.

However, several hosts reported not having sufficient time for one-on-one interactions with their guest speakers. One host indicated that her speaker had project deadlines to meet during the time of her visit to campus, which limited the speaker's availability for interactions other than the lecture. Another host commented that it is important to ensure that the host's department head is aware that the program is intended to primarily benefit the host rather than the department. As a result of this feedback, we will encourage prospective hosts to ensure that the travel

TABLE 2. Activities of ADLS Speakers, as Reported by Hosts

Activity	Number Reporting
Presented research seminar	18
Engaged in one-on-one interaction with host	18
Met with students in the host's department or college	18
Met with faculty members in the department or college	16
Participated in hands-on activities, such as teaching and experiments	13
Discussed future collaboration with host	13
Interacted with individuals working in host's laboratory	12
Discussed experimental approaches, data analysis, and research projects	9
Presented a guest lecture in one of the host's classes	4
Discuss funding opportunities and grant proposal writing with host	4
Shared a meal with students in host's research program	3

schedule of the speaker includes sufficient time for individual discussions as well as formal meetings with others. We also will use this and other feedback from the hosts to develop a brochure containing tips for new faculty members who are considering hosting speakers and with guidelines for department heads.

Table 3a summarizes the benefits reported by hosts regarding their interactions with their invited speakers. Because there was a wide range of benefits reported, a specific benefit was typically reported by only one or a few hosts. The more frequently mentioned benefits include establishing a relationship with a nationally recognized individual who could serve as a mentor; obtaining feedback on current research projects and exploring ideas for future research projects; having the opportunity to discuss various collaborations with the speaker, including writing papers and research proposals; and receiving suggestions to improve teaching.

Table 3b provides a list of types of information that hosts gained from discussion with their speakers. We have grouped similar responses from the hosts into more general categories, with categories listed in order of decreasing frequency.

One host faculty member reported that the seminar and the speaker's discussions with her colleagues excited them about the potential of her

TABLE 3a.   Benefits of Hosting the Speaker

Gained important professional contact
Obtained feedback on and suggestions for future research proposals
Discussed collaborating with the speaker
Gained experience that will be beneficial in teaching
Increased networking within professional organizations
Obtained experience in hosting a speaker
Networked with on-campus faculty as a result of the speaker's visit
Speaker recommended students for host's Ph.D. program
Received advice on design of laboratory facility
Speaker will serve as a reference person for the host
Learned new statistical techniques
Identified scientists in related fields for interdisciplinary projects

TABLE 3b.   Useful Information Provided by Speaker

Literature references for research (books, journals, etc.)
Research protocols
Procedures used at speaker's research laboratory
Companies from which to order specialized laboratory equipment
Teaching tips
Clinical experience

research specialty and resulted in a number of her on-campus colleagues considering possible collaborations with her. This host capitalized on this interest by organizing and conducting a weeklong on-campus workshop for her colleagues who were interested in this area of research. The workshop provided the opportunity for these faculty members to develop molecular biological reagents that will be used to initiate new collaborative research projects with the ADLS host. Another outcome has been that one of the host's colleagues is auditing a new graduate-level course she is teaching. These activities illustrate the kinds of secondary outcomes that can accrue from a single seminar visit. We speculate that these outcomes will lead to additional benefits for the ADLS host in the future. These benefits may include creation of a critical cluster of researchers with common interests who can clearly articulate the significance of the host's research and serve as advocates when she becomes eligible for tenure.

Another host indicated that her speaker, a clinician, led rounds and described unusual cases for a number of veterinary medical residents. Two women reported hands-on interaction with their guests in learning new techniques, one in the area of teaching and the other in the area of research. All of the hosts noted that the speakers met with students in the department, as well as with students working on the host's research projects.

In addition to benefitting from the on-campus visit of the speaker, host women faculty members also are continuing to interact with their invited guests in a variety of ways. These are summarized in table 4 and

TABLE 4.   Continued Interactions with the Speaker

Activity	Number Reporting
Plan future professional collaboration	11
Discuss ongoing and future research	8
Meet at professional conferences in the future	5
Visit by host to the speaker's university	5
Introduce host to speaker's collaborators	4
Exchange research reagents	2
Submitted paper to journal edited by speaker	2
Review host's papers and grant proposals	1
Invite speaker to present at K-State again	1
Provide funding for host's research	1
Develop graduate certificate program	1
Propose K-State as site for professional society meeting	1

include planning for future meetings and collaborations, writing joint grant proposals, preparing and submitting publications, and arranging reciprocal visits to the speaker's institution.

Some of the speakers also have provided letters of support for grant proposals, nominated the host women to serve on committees of national professional organizations, introduced them to other professional colleagues, and recommended graduate students to work with the host woman faculty member at K-State. One of the speakers plans to return to K-State to present a more extensive hands-on clinic for students. Five of the women mention plans to meet with their speakers at upcoming national professional conferences. These planned meetings will create opportunities for the women to be introduced by their guest speakers to senior leaders in their fields. Such meetings will be potentially catalytic for a rapid expansion of the women's professional networks.

Some respondents have already described such professional connections made as a result of the program, including meeting influential individuals in professional organizations, companies, and other universities. Table 5 summarizes the types of contacts that hosts have made to date.

Challenges

Because of the extensive and varied benefits that our hosts have reported, our desire is to have every eligible woman faculty member participate in the ADLS program. However, we have not yet achieved that goal. In early fall 2004, we hosted a lunch meeting with STEM women faculty members to share information about the K-State ADVANCE project-level initiatives. At that meeting we learned that some of the junior women faculty members were reluctant to apply and participate in the ADLS program because they were hesitant to correspond with well-known individuals in their discipline. We expect that this concern may be partially responsible for the fact that not all eligible women participated in the first two years. Two other factors that may

TABLE 5.   Contacts Made That Will Benefit Host's Career

Individuals in professional organizations and companies
Journal editorial board members
Researchers at speaker's institution during reciprocal visits by host
Faculty in the same research discipline at other institutions

contribute to some junior faculty members not participating are the lack of encouragement by their department heads to take advantage of this opportunity, and the feeling of being overwhelmed by the teaching, research, and service duties associated with their new position.

Another challenge has been poor attendance at a few of the lectures. At least three reasons may have been responsible: conflicting events, inadequate advertising, and departmental culture. One of our hosts is in a department with a primary focus on undergraduate teaching and a limited emphasis on research and has found it difficult to engage her colleagues in her speakers' visits.

We are addressing these challenges by (1) working with department heads to be advocates for this initiative and to take a proactive role in promoting this opportunity for eligible women in their units; (2) asking the ADVANCE Steering Committee members from each of the four colleges to contact the eligible women in their college, encourage them to participate, answer any questions they may have, and assist them, if necessary, in identifying appropriate speakers; (3) meeting with groups of STEM women faculty to share the experiences of those who have already hosted speakers and the benefits that they received from their participation; and (4) working more closely with individual hosts who experience difficulties with a visit to address the concerns and improve the scheduling and implementation of future visits.

Future Assessment Plans

We plan to continue to collect follow-up information from the host faculty members to assess the longer-term impact of the ADLS program. Specifically, we will be following up with the women each year until their tenure decision to determine the impact that hosting these speakers has had on their careers, whether the speakers are serving as outside references as part of their tenure package, and what additional interactions with their speakers have taken place. We also plan to request information from the speakers as to the benefits they have derived from participation in this program and any suggestions they may have for future program directions.

Expansion and Institutionalization

The ADLS program has fostered synergistic interactions among the ADVANCE project and other university lecture series. The provost has included an ADVANCE Distinguished Lecturer as part of the Provost's Lecture Series and has provided financial support for other university-wide speakers that the ADVANCE program has hosted. Both the Johnson Center for Basic Cancer Research and the Department of Chemistry have cohosted with ADLS the presentations by two leading women scientists invited to speak on our campus. We expect this type of collaboration to expand to other campus lecture series.

We will be seeking financial support from individual donors and corporations to sustain this program in its present form for STEM junior women faculty members beyond the NSF grant period. However, it is our goal to encourage each STEM department to adopt this lecture series model as a part of the culture of the unit in promoting the success of all junior faculty members. We recognize that departmental resources may not be sufficient to allow each junior faculty member to host a seminar speaker each year during the probationary period. Nevertheless, we want to encourage departments to give junior faculty members priority for issuing seminar invitations to guest speakers who could become important members of the junior faculty member's professional network.

We also want to share this model with departments on our campus outside of STEM disciplines and with other universities. The concept of using the vehicle of a lecture series to develop the professional network of a junior faculty member is easily transferrable to other disciplines and other institutions. We believe that participation in this initiative will accelerate the development of networks for our new faculty members; include women in decision-making roles; create links between women faculty and disciplinary leaders; and lead to the increased likelihood of tenure and promotion of the women. As we compile additional data on the benefits and success accrued as a result of this program, we will be in a position to make a strong case to college deans and the provost to work with the university foundation to obtain sufficient funding so that all junior faculty members can fully participate in this program.

Conclusions

We have created a unique seminar series that allows tenure-track women faculty to identify a leader (male or female) in their research areas to invite to campus. It offers women the chance to showcase their research programs to disciplinary leaders who may in the future serve as external evaluators of their grant proposals, journal articles, or tenure packages. The campus visits feature one-on-one time for the faculty member with the speaker, and when possible, visits by the speaker to the faculty member's classes, and participation in other activities. This program also raises the visibility of the host with her departmental colleagues and may lead to increased interactions with them. Such interactions can lead to her colleagues being more knowledgeable about her research and thus stronger advocates for her during the tenure process.

Junior faculty women from all four STEM colleges that are part of our ADVANCE project have served as hosts during the first two years of this program. In the reports they have provided and through their interviews and conversations with the ADVANCE project leaders, these women expressed their enthusiasm for their experiences and their interactions with their guests. However, not all of the eligible women have invited speakers, and thus, it will continue to be very important for senior faculty members, department heads, and other administrators to reinforce the importance of participation and provide assistance in the selection and invitation process.

While this program is in the early stages of its implementation, and we have data from only two years' participation to report, this initiative is already producing positive outcomes. We are particularly encouraged that the women who have participated are selecting very prominent individuals in their fields to invite to campus for this seminar series. Our junior STEM faculty women may not have considered hosting such prestigious guests without the impetus of the ADLS opportunity. Furthermore, these distinguished guests might not have come to our campus otherwise. In fact, the host of a recent speaker commented, "He seemed to really enjoy the visit and I had great interactions with him. It is definitely valuable to host big people in my field, and in this case I know he never would have come had it not been for ADVANCE. So I feel fortunate I was able to host him."

All members of the host's department benefit from the opportunity to interact with such leaders, which reduces the isolation produced by our geographic location. We have found that several of the visits have resulted in establishment of research collaborations between the K-State faculty members and their guests, an outcome that is even more positive than we anticipated.

One of the outcomes we hope to have from this program is to create a climate in which the norm is nurturing junior faculty members and aiding them in the establishment of their professional networks. A simple way to do this is by giving them preference in selecting and hosting speakers as part of the regular departmental seminar series.

The ADLS is just one of a variety of programs in our ADVANCE project that facilitates the development of faculty networks. Other programs provide support for professional travel to conferences at which the junior faculty may present a paper or renew contacts with ADLS speakers and travel to pursue collaborative research with colleagues at other institutions. Group and individual mentoring programs are also components of our ADVANCE project. These programs have the potential to be synergistic in building the professional contacts that have been shown to be so valuable in contributing to the success of all faculty members (CAWMSET 2000; Pattatucci 1998; Sonnert 1995; Trower 2002).

We look forward to following the careers of the women who participate in the ADLS program and our other ADVANCE initiatives and observing the growth of their professional networks. We predict that involvement in these networks will significantly affect their success.

NOTE

We want to thank the Kansas State University (K-State) ADVANCE Project Executive Committee for their support and participation in this initiative. They helped formulate the guidelines for the ADVANCE Distinguished Lecture Series program, have been instrumental in encouraging eligible women faculty members in their colleges to take advantage of this program, and have funded and attended the dinners for each speaker and the junior faculty host. We also wish to thank the K-State ADVANCE Project Steering Committee members for their assistance in reviewing and approving the proposals for the speakers and the visits. We are grateful to the K-State Office of the Provost, the colleges, and the departments for the supplemental funding they have provided to support the expenses associated with the program. The Office of Educational Innovation and Evaluation has administered evaluation instruments and conducted interviews with program participants and provided the Executive Committee with their findings. Finally, we thank Ms. Rebecca Wood, K-State ADVANCE project coordinator, who has been responsible for communicating with eligible faculty members about the program, advertising the lectures to the campus community, collecting information about the speakers and their visits, and helping prepare the information presented in this chapter.

The K-State ADVANCE Project is supported by the National Science Foundation under Cooperative Agreement number SBE-0244984. Any opinions, findings, and conclusions or recommendations expressed in this material are those of the authors and do not necessarily reflect the views of the National Science Foundation.

REFERENCES

Commission on the Advancement of Women and Minorities in Science, Engineering, and Technology Development (CAWMSET). 2000. *Land of plenty: Diversity as America's competitive edge in science, engineering, and technology.* Washington, DC: U.S. Government Printing Office.

Etzkowitz, H., C. Kemelgor, and B. Uzzi. 2000. *Athena unbound.* Cambridge: Cambridge University Press.

Kansas State University (K-State) ADVANCE Project. 2005. ADVANCE Distinguished Lecture Series. http://www.ksu.edu/advance/SeminarsEvents/distinguished_series.htm. Accessed June 30, 2005.

Kansas State University (K-State) ADVANCE Project Indicator Data. 2005. ADVANCE Institutional Transformation Grant National Science Foundation Indicators, 2004–2005. http://www.ksu.edu/advance/K-State%20Gender%20Data/Condensed-NSF-Report-of-Indicators-2005.pdf. Accessed June 30, 2005.

Pattatucci, A. 1998. *Women in science: Meeting career challenges,* Thousand Oaks, CA: Sage.

Sonnert, G. 1995. *Who succeeds in science? The gender dimension.* New Brunswick, NJ: Rutgers University Press.

Trower, C. A. 2002. Women without tenure, part 3: Why they leave. http://nextwave.sciencemag.org/cgi/content/full/2002/03/18/3. Accessed June 30, 2005.

FIVE

Interconnected Networks for Advancement in Science and Engineering

THEORY, PRACTICES, AND IMPLEMENTATION

Mary Lynn Realff, Carol Colatrella, & Mary Frank Fox

INTERACTION AND exchange are critical to performance in science and engineering fields because scientific work revolves on the cooperation of people and groups, and requires human and material resources (Fox 1991, 2000, 2001, 2003; Pelz and Andrews 1976). Sciences are highly interdependent enterprises, involving the combination of people and skills (Bradley 1982) and access to shared resources (Wray 2002). Collaborative (coauthored) work fares better in the publication process than does solo research, in part because it is subject to "checks" by multiple persons (Presser 1980), and collaborative work has greater "impact" in science (Beaver 2004).

Women in science and engineering disciplines are disadvantaged in resources and rewards to the extent that they are "outside" of networks of interaction and exchange (Fox 1991, 1996, 2001). Although women faculty members are de facto members of departments and disciplines in universities, women are not always automatically included in informal mentoring or research-based networks (Hornig 2003; Rosser 2004). Georgia Tech (GT) ADVANCE has responded to this need of women faculty in science and engineering environments by expanding and developing various ADVANCE networks.

Networking has become a principle of modern work in many fields because it allows those who would be excluded to overcome exclusion, connects strivers to role models, and fosters mentoring. Business journalists identify the glass ceiling for women leaders, who are less likely to achieve positions as chief officers. Most recently, *The Economist* published a special report noting mechanisms that could increase the number of women executives: forming women's networks to supplement or supplant "male clubbishness," combating existing male bias against promoting women, and spotlighting role models.[1] As Joann Moody writes, "To achieve financial and professional advancement, professionals must (almost always) become networked with a variety of powerful people who can open doors of opportunity, help them solve problems, alert them to impending dangers and reorganizations, and introduce them to yet more power-holders" (2004, 43).

In the past twenty or so years especially, women faculty members in engineering and science have worked to establish disciplinary networks that could connect women in different institutions working within related research fields. Given low numbers of women in engineering and physical sciences, many women faculty rely on such groups as a means of establishing collegial relationships beyond institutional structures. For example, recognizing the value of connecting women in similar fields with one another, NSF ADVANCE funding enabled the creation of the Women Engineering Faculty Leadership Network, which sponsors several meetings each year for women engineering faculty at universities across the United States.[2] This network concentrates on bringing together women faculty who might be in the minority in their home institutions and on teaching them "transformative leadership techniques." Such disciplinary networks enable individuals across institutions to develop a broad understanding of barriers and facilitators for women in specific professions, and encourage their members' career progress via mentoring and skills-building workshops, while sometimes sponsoring awards and providing opportunities for research collaboration. An institutionally based system of interlocking networks such as those at Georgia Tech, some disciplinary and some not, can be a critical mechanism for any university seeking to change cultural attitudes impeding access and equity for women and other minorities.

GT ADVANCE Means for Creating Networks

Enhancing Existing Networks

Universities already include groups that function as networking opportunities, such as committees focusing on hiring, evaluation, and curriculum or departmental research seminars. In these groups, faculty may interact with members of their own or other departments and become better acculturated to their academic environments. Research centers at universities constitute another type of network, allowing faculty with common interests to share ideas and to foster collaborative research projects. Yet these existing infrastructural groups may not include women faculty, and their purview does not ordinarily include the issues of climate, decision-making, and advancement that women faculty acknowledge as being significant for their long-term satisfaction with their work and institutions (Fox and Colatrella 2006).

An example of a network focusing on representation issues, the Georgia Tech Center for the Study of Women, Science and Technology (WST) (1998–present) played a critical role in designing the ADVANCE grant proposal as an existing network of women faculty interested in promoting women's advancement on campus. The WST Center connected faculty and administrators across colleges at Georgia Tech who expressed interest in pedagogies, research, and professional issues related to the underrepresentation of women and minorities in science and engineering; in leadership strategies in relation to advancement; and in theoretical models in gender studies.[3] The WST Center partnered with College of Engineering and the provost in developing the GT NSF ADVANCE program. The Center continues under the shared direction of Mary Frank Fox, Carol Colatrella, and Mary Lynn Realff, and sponsors activities and partnerships that support the full professional participation of women on campus, thereby extending the scope of its network to include all those interested in improving the academic climate for women faculty.

Establishing New Networks

The GT NSF ADVANCE team, composed of the PI (principal investigator) and co-PIs, the program director, coordinator, ADVANCE professors, and WST codirectors, meets biweekly. The ADVANCE team disseminates information about activities via campus news media as well as through the ADVANCE web site and the annual report. Mentoring occurs in a variety of settings: one-on-one sessions, group discussions, and the college networks coordinated by ADVANCE professors (described in detail in the next section); WST faculty-student mentoring pairs, and WST-ADVANCE workshops that include sessions on developing grants, proposal writing, and promotion and tenure.

College networks developed by ADVANCE professors and cross-college activities established with WST ensure that all women faculty and any interested male faculty benefit from the initiative. To further expand the existing WST network and to include administrators, GT NSF ADVANCE established a range of communication and mentoring networks. Some networks such as the WST ADVANCE network affect the entire campus community, while others connect individual faculty members and institute offices working on particular issues such as evaluation (Promotion and Tenure ADVANCE Committee, the minority faculty network coordinated by Faculty Support Services), institutional policy development (Faculty Senate, Executive Board, ADVANCE project team), or curriculum development, or groups of faculty in particular schools or colleges interested in university topics or, occasionally, issues of balancing family and work (Computing Mothers Group). While there is not a connecting framework that brings various groups together under one aegis, the ADVANCE team shares information about groups on a regular basis in order to direct individuals to appropriate mechanisms.

WST-ADVANCE, college, and intercollege activities have brought together women faculty in small groups based on research interests, college location, and personal interest in leadership topics. These formal and informal networks have encouraged women to put themselves forward for advancement, to establish research collaborations, and to communicate more effectively with college and institutional administration. Networking within an organization provides individuals with opportunities to share information and wisdom, while the organization benefits from this leadership.[4] The ADVANCE team and faculty participating in ADVANCE activities have established better lines of communication with each other and with central administration. ADVANCE has consequently served as a means of advancing women into administrative appointments and promoting them into leadership positions.

Management consultants recognize the ability to "share successes" as a critical feature of leadership that improves the likelihood corporate workers will succeed in management positions (Kanter 1977, 185). Networks built across disciplines and interest groups similarly benefit universities by promoting organizational change because conversation and debate among faculty and administrators address problems that might elude or stymie more parochial groups. GT ADVANCE has concentrated particularly on establishing clear, consistent guidelines regarding evaluation procedures and on addressing the campus climate as it affects women faculty. ADVANCE conference breakout groups have brought together senior administrators (mostly male) and women faculty together in discussions to share standards, practices, and procedures and to develop improvements. Interactions among women faculty, and with senior, typically male administrators, are opportunities to effect changes in attitudes and culture as well as policies and procedures (including "gender schemas," Valian 1999) that "reframe" or "correct" forms of stereotypes (Jamieson 1995, 190).

Networks of minority group members within a larger organization allow members of the minority group to support each other and to cat-alyze collective action for improvement. Members of majority groups also benefit from the development of networks connecting minority group members, if the minority networks are linked into other groups within the university. Building and maintaining such networks enhances majority understanding of the barriers and facilitators for women faculty and improves opportunities for collaboration, including collaboration in research. Georgia Tech's ADVANCE research suggests that weaving networks involving women into existing and new university infrastruc-ture can improve women's participation in institutional decision-mak-ing and their prospects for advancement. Since GT ADVANCE began, there has been a slight increase in the number of women promoted to full professor (nineteeen promoted to full professor from 2000 to 2005) and in those appointed to chaired professorships, unit chair positions (2), positions as associate dean (2), and directors of institute centers (1).

ADVANCE Professorships

At the outset of designing the Georgia Tech NSF ADVANCE project, faculty and administrators recognized the need to create mechanisms of change that would be sensitive to the needs of faculty at disciplinary, unit, and college levels and that would offer a means of mediating between faculty and administrators. The institution committed to create an intercollege network of termed professorships and endowed chairs. During the grant period the NSF award has been used to support the educational, research, and ADVANCE activities of four senior women faculty, one in each of the participating Colleges (Computing, Engi-neering, Sciences, and Ivan Allen College of Liberal Arts). Each profes-sor receives a budget for college ADVANCE activities and a fund to support advancement of her research program. The supported faculty members commit to activities that develop and sustain the community and advancement of women faculty in their colleges, and in turn, across colleges. The ADVANCE professor network nurtures and exemplifies the advancement of women as an issue at the center of Georgia Tech.

Activities of each ADVANCE professor focus on developing college and cross-college networks to effect institutional transformation, including (1) developing a network of women faculty, committed to institutional transformation; (2) advising the deans of respective colleges and the provost on issues vital for the success of women faculty; and (3) facilitating and organizing annual retreats for women faculty and administrators to consider barriers and facilitators affecting women's advancement. To support the ADVANCE professor network, Georgia Tech has funded a program director and coordinator, who furnish and keep current a database of information to be disseminated, and provide logistics, communication, and administrative assistance.

The recipients of ADVANCE professorships were chosen using criteria of full professorial rank, national and international stature, achievement and impact in research and education, and demonstrated commitment to advancing women in science and engineering. Georgia Tech has committed to sustaining these supported positions beyond the period of the grant by finding support through donors for professorships or chairs, or by sustaining them through institutional funding. The support will be at the same level as that provided through the NSF award. These prestigious professorships are termed (five years). They may be renewed or may be used for support of others over time.

Any university can implement a network of ADVANCE-type professors to foster institutional change, recognizing that the extensive time commitment requires a substantive budget to support faculty activities and to allow research to continue unconstrained by such commitments. Mentoring and other networks devised to assist individuals and change the climate must be a joint responsibility of the institution and interested leaders. That is, transforming an institution should not only be the responsibility of individual change-agents but also of the institution, more broadly and collectively. Faculty serving as change agents ought to have appropriate authority, compensation, and recognition and ought to be provided with suitable resources.

Mentoring Networks

Women in colleges with ADVANCE professors are members of the recently developed college networks. Each ADVANCE professor began her duties by gathering information about needs and desired outcomes from women faculty in her college. Information gathering included focus groups and meetings, surveys, and one-on-one dialogues; each professor was expected to assess needs, but different college populations suggested different means of assessment. Assessing the implications of data collected by these means as well as institutional data and findings from the Georgia Tech ADVANCE research program's survey of institute faculty provided each ADVANCE professor with the opportunity to fine-tune programs developed for the needs and interests of faculty in her college. Biweekly team ADVANCE meetings that bring together the ADVANCE professors with the grant leadership team allow best practices to be shared and adapted to particular disciplinary concerns.

For example, ADVANCE professor Jane Ammons noted that many women faculty in engineering had not previously interacted with each other and that women faculty could benefit from individual mentoring regarding advancement. Ammons has hosted lunches to bring together the College of Engineering's female faculty in configurations to foster research collaborations. She has also developed a method of "speed mentoring," a career coaching practice that allows a junior faculty member to receive advice regarding her vita and career building from different mentors during one session. Ivan Allen College ADVANCE professor Mary Frank Fox understood that IAC women faculty believed that their professional accomplishments were not adequately shared with colleagues; they also asked for assistance in developing grants. Fox has developed a web site collating liberal arts professors' research interests and accomplishments, which serves as a way of articulating disciplinary achievements in an interdisciplinary format. She has also hosted a series of workshops devoted to advancement, and cosponsored an annual grants workshop with the Ivan Allen College associate dean for research. Noticing the need for faculty mentoring, Professor Mei-Yin Chou has coordinated opportunities for College of Science female faculty to meet with successful women scientists and has developed, under the coordination of a dean's staff member, an extensive mentoring program that includes pairing senior and junior faculty and putting advancement resources on the web. Mary Jean Harrold, ADVANCE professor of computing, built on her experience with national organizations in computing to respond to her faculty's needs. Harrold has worked with female faculty and administrators in her college and in the newly developed National Center for Women in Information Technology to identify the particular barriers for women in computing that inhibit growth in IT fields (see http://www.ncwit.org). Such barriers include coping with the low numbers of women students, which the College of Computing has addressed by hiring an assistant dean for special programs who works on recruitment and retention, and encouraging women faculty who work in different subdisciplines to meet and mentor each other. Although Georgia Tech has a high number of women faculty in computing relative to national averages, they are still a distinct minority within their college.

Research Foundations of Networks

The GT ADVANCE professor college networks function as network hubs that intersect with other local and national hubs. In addition to hosting college-specific events, the ADVANCE professors work with the ADVANCE program director and coordinator and codirectors of the WST Center to develop cross-college events and workshops for faculty devoted to understanding strategies for advancement, family-friendly practices, career coaching, and building one's skills for effective leadership. The ADVANCE team has also sponsored workshops for school chairs on best practices in recruitment and retention. Bringing together women faculty according to their fields within a college is a first step in forming bonds among those likely to have similar or complementary research interests, which is an important factor for faculty in research universities. ADVANCE research surveys conducted by Mary Frank Fox indicate that many women faculty do not speak frequently with colleagues in their units about their research. For example, in significant contrast with their male counterparts, 30% of whom speak to other faculty in their home units about their research on a daily basis, 13% of the women faculty reported speaking to faculty in their home units about their research on a daily basis.[5] Other institutions might identify a different set of common needs and interests to be tapped in forming networking groups.

Over time, college-specific events have been complemented by cross-college WST ADVANCE activities. Events sponsored by WST ADVANCE serve as opportunities for faculty to communicate about their research and related leadership issues within and beyond networks. Clarifying processes for grant writing and advancement, as well as addressing work-family issues and other institutional policies, cross-college WST ADVANCE workshops, open to all faculty and targeted to female faculty, have further strengthened networks of faculty women across disciplines, which in turn have effected institutional change.

For example, a recent workshop on interdisciplinary grants included representatives from the Office of Contracts and Grants and the vice provost for research, and provided the opportunity for faculty women to refine their understanding of institutional procedures for interdisciplinary, competitive grant proposals. Noting that the administrators depended upon their personal knowledge of faculty research areas to suggest collaborators for grants, faculty women present at the workshop expressed concern that they were not always "in the loop" on such emerging proposals and recommended that the university develop more transparent means of informing faculty about collaborative projects in formation. In recommending a web site listing grants in the process of development and internal competitions for grants with limited submissions, women faculty acknowledged that they can be outside existing informal communication networks and noted that a web site would benefit all faculty members. Having such a web site listing developing grant opportunities makes the process of finding potential collaborators more open and potentially available to all faculty members.

Many cross-college WST ADVANCE events coordinated by the ADVANCE team and the WST codirectors involve faculty from different units who share perspectives and personal narratives on career pathways, advancement, and research productivity. A panel on building campus community life included two administrators, two faculty, and two students and provided continuing input on problems identified at an earlier ADVANCE conference, such as safety, stress, unit boundaries inhibiting research collaboration, and the difficulty of resolving issues that cut across the reporting lines of academic and student affairs. Follow-up reports on these issues were addressed at later ADVANCE conferences and publications disseminated across campus. At another event, a recently hired female endowed chair and codirector of a research center talked about her career path in a newly developed field and offered advice for the thirty-five faculty women from different units concerning how to establish a repertoire of skills and strategies that would flexibly meet career demands and how to develop students and colleagues to be partners in one's career success. A panel of administrators and faculty later extended this conversation to include tips for becoming an administrator.

How Networking Functions to Advance Faculty

Critical to the success of GT NSF ADVANCE is the overlapping, interlocking relationship among the existing and newly formed networks. A flow chart of the networks in place and the lines of interaction among individuals participating in them would not show neat divisions of authority. Instead of a chain-of-command structure, what has emerged is a management system reflecting the many relationships that each woman faculty member might tap: connections with unit and college colleagues, alliances with women in other units and colleges, and gains from participating in workshops and conferences with faculty and administrators throughout the university. For example, individual women have turned to the ADVANCE professors to provide advice and support in negotiations with school chairs regarding resources and teaching load. Supported by their college deans and the provost, ADVANCE professors can advocate for current faculty and can recommend candidates to be considered for prospective appointments to school chairs.

Networking and related initiatives promoting advancement have transformed cultural expectations and realities of women's success, as demonstrated by the increased numbers of women faculty being promoted and tenured to associate rank, those attaining the rank of full professor, and those appointed to significant leadership roles as school chairs and associate deans. During workshop discussions and in research surveys and interviews, an additional number of women faculty members have recently expressed interest in becoming administrators. Participating in networks has made women faculty more aware of career opportunities and provided greater access to information necessary for advancement. Two of four ADVANCE professors have recently been promoted into administrative, leadership positions, one as associate dean and one as a school chair.[6]

The Georgia Tech NSF ADVANCE professors and the program as a whole have developed particular networks around research, including the WST network of researchers, connecting women faculty members with each other as well as with administrators. The ADVANCE program team coordinated by the program director includes the PI, the co-PIs, the ADVANCE professors, and other contributors.[7] As noted above, the ADVANCE team meets biweekly throughout the year to track progress on research and other initiatives, to plan the annual con-

ference, to share best practices, and to do long-term planning for institutionalization of ADVANCE initiatives. Because the team is diverse
and represents central administration and faculty from different colleges,
their collaboration works to disseminate ADVANCE projects throughout the university. Faculty members on the team manage the grant,
budget, and conference; coordinate the research initiatives, and disseminate information about activities. Administrators on the team, the
provost and a dean, understand how to gain access to necessary human
and material resources in order to effect change. Weaving research, theory, and practice into pragmatic policy development, the ADVANCE
team serves as a model network for other universities—demonstrating
the necessity for effective networks to include participants from different organizational levels. Its deliberations are critical to the functioning
of the program and its continuing success.

Similarly, the intercollege network of ADVANCE professorships
promotes the goals of ADVANCE to enhance the retention and
advancement of women faculty. The ADVANCE professors have built
networks of communication, mentoring, and exchange among female
faculty in their colleges (from 2001 to 2006 in Computing, Engineering,
Science, and the Ivan Allen College of Liberal Arts).[8] Each ADVANCE
professor serves as a role model and mentor for other women faculty and
as a liaison between the women faculty in her college and college
administrators. When appropriate, the professors consult with their
deans to define existing constraints and to develop strategies enabling
the advancement of women. Each ADVANCE professor regularly
meets with the women faculty in her college and sponsors college activities. ADVANCE professors share best practices and collaborate on
cross-college workshops on family-friendly policies, grant and proposal
writing, leadership skills, and experience from successful women. Hearing about how particular female colleagues have navigated difficulties to
successful resolution inspires workshop participants to manage their
careers and to prepare for opportunities for advancement.

The GT NSF ADVANCE conference brings together junior and
senior women faculty with administrators from all areas and levels of the
institution. Four GT conferences (2002, 2003, 2005, 2006) and one
national conference (2004) have been held. Approximately 120 faculty
members attended each GT conference, while the national meeting
attracted more than 250. At these conferences, faculty, provosts, deans,
and school chairs review and refine goals and progress of the program.
The program director's annual presentation provides updates on the

issues raised by conference participants and what initiatives have been undertaken in response to concerns. The research coordinator presents data from surveys of Georgia Tech and benchmark institutions. The PI and co-PIs report on institutional development and collaborate with the External Advisory Board members to consider long-term prospects for Georgia Tech and to discuss GT ADVANCE initiatives in the context of national efforts in these areas.

As advancement is the focus of the GT ADVANCE program, promotion and tenure policies, best practices, and instruments to eliminate bias in evaluation procedures are shared at the annual conference as well as at other workshops. GT NSF ADVANCE has worked to institutionalize a formal training process for committees involved in tenure and promotion, as addressed in chapter 11 in this volume, "Equity in Tenure and Promotion: An Integrated Institutional Approach."[9] The range of units and practices represented by members of the Promotion and Tenure ADVANCE Committee constitutes another important aspect of ADVANCE networks, as the committee's report and workshops based on its work have encouraged faculty to discuss evaluation practices, procedures, and expectations.

The annual conference also includes opportunities for female faculty to converse with senior colleagues in different units and administrators who can advise about career advancement. Conference conversations between women faculty and administrators include discussion of evaluation procedures and expectations. Based on a model developed by the College of Engineering ADVANCE professor, the most recent conference included a curriculum vitae review session in which faculty were paired so that one senior faculty member met with a junior faculty member from another unit to discuss career progress and strategies for improvement. This popular session allowed faculty women to benefit from "insider" information in a nonconfrontational format and offered a structured—and open, equitable—means of meeting a senior faculty member who might continue to offer mentoring and advice.

Recent testimonials bear witness to the significance of mentoring accomplished through the ADVANCE professor network. One female faculty member commented, "I have immensely benefited from the series of women engineering faculty lunches [she] has organized in her role as the NSF ADVANCE Chair. During these lunches, not only was I able to meet other women engineering faculty and learn from their experiences on important issues such as balancing life and career, [and] have senior engineering faculty critique my resume, but also was able to meet and interact with higher level administrators (for example the dean of engineering) in an informal setting."

Another individual commended the way the ADVANCE professor in her college was able to identify resources to support graduate student travel when ordinary means of funding were not available. Other female faculty members report that being able to consult with their ADVANCE professor has enabled them to navigate the complex bureaucracy of a research university. One junior faculty member notes that she appreciated receiving such advice, as her department "is often a labyrinth in terms of knowing how things 'actually get done,' understanding the politics . . . of the place and what are the rules and procedures." To adopt the language of another affected faculty member, this mentoring "operationalizes" the ADVANCE project's principles to advance women faculty

ADVANCE professors are able to advise individual faculty members about specific career choices and life circumstances in one-on-one meetings and lunch workshops. A key finding of GT ADVANCE research is that women often lack knowledge of resources and how to obtain them. The ADVANCE networks established by the professors provide information about rules and procedures and offer one-on-one support and career guidance for the work and life choices of individual women faculty, serving as a potential model for other universities.

Tracking Equity and Enhancing the Environment

Georgia Tech NSF ADVANCE has relied on central network resources to collect and track institutional-level data related to gender equity and to provide information about newly developed family-friendly practices. The central offices of Institutional Research and Planning (IRP) and Assessment, and their directors, contribute to ADVANCE initiatives and assessment along with the external evaluators. Conferences have been evaluated by participants, using instruments developed and analyzed by Assessment. IRP generated institutional outcomes data as required by NSF that measures faculty compensation, evaluation, advancement, and resource allocation.[10]

Tracking institutional data and sharing that information in institute-level groups has enabled administrators and faculty to understand the specific conditions of the university environment for all faculty, particularly for women. In addition to sharing data analyses regarding women faculty, having the ADVANCE program director and co-PIs involved in policy groups raises awareness of specific barriers and facilitators for the success of women faculty, especially for those on campus not directly involved with the project. For example, this past year's deliberations of the campus diversity council included participation from ADVANCE team members who were responsible for providing a subreport describing current demographics and five-year trends concerning faculty diversity. Thus, ADVANCE research and suggested best practices guided the subreport recommendations.

Georgia Tech's administration has contributed to the success of ADVANCE by working with the state Board of Regents and with networks of interested faculty to create family-friendly practices. Participation of ADVANCE leadership has also helped to strengthen and extend the scope and impact of these practices. ADVANCE team members advised on the creation of the Active Service Modified Duties process and have consulted with the staff officer coordinating the office of Faculty Support Services. The Active Service Modified Duties process allows faculty affected by health, pregnancy, or caregiving responsibilities to apply for release from particular teaching or scholarly duties and to ask to stop the tenure clock. ADVANCE advisors have consulted on the building of lactation facilities for nursing moms. Partnering with a professional child-care provider, Georgia Tech helped to develop a daycare center that has helped both male and female faculty balance their commitment to family and work. These policies and practices were created through networking, and their implementation has benefited from ADVANCE networking as well. An institutional cultural change regarding balancing family and work has occurred because these policies and facilities are publicized among faculty, sometimes in WST ADVANCE workshops, and they have already been used by a number of faculty members who endorse their development. In this way, family-friendly policies and facilities have become normalized as part of the environment through their development and dissemination in networks.

Conclusion and Recommendations for Implementing Networks for Change

In its "integrated institutional approach" to factors that support the advancement of faculty and provide a model of best practices in academic science and engineering, the GT ADVANCE program is research based. The program emphasizes organizational factors and features that are shown to shape positive outcomes of participation and performance among faculty in science and engineering. Accordingly, both prior research about the importance of collegial and collaborative networks of access and opportunity, and the findings of the GT research ADVANCE program, underlie the networks created and sustained through the GT ADVANCE program. It has been easy to establish networks among individuals who are grateful to find supportive connections but more difficult to develop and understand their mechanisms of power. All evidence of any change from developing networks is only anecdotal at this stage, although we anticipate seeing some permanent shift after analyzing the 2005–6 ADVANCE research survey.

As described in this chapter, these networks include (1) the expanded WST Center networks; (2) new networks established through the GT ADVANCE program team, the ADVANCE professors, and the GT ADVANCE Tenure and Promotion Committee; and (3) interconnected mentoring networks of individual change-agents such as the ADVANCE professors as well as broad, collective networks created through ADVANCE initiatives such as the university's interdisciplinary grants workshop and the national network of investigators coordinating NSF ADVANCE Institutional Transformation grants.

Current plans for institutionalizing gains made by ADVANCE at Georgia Tech include sustaining workshops through the Center for the Study of Women, Science, and Technology and monitoring family-friendly policies and facilities through the Office of the Provost. Tracking of institutional data will be ongoing as coordinated by Institutional Research and Planning and working with the Office of the Provost. Development efforts are in progress to secure funding to continue and supplement the ADVANCE professors and their contributions to networking.

The success of Georgia Tech's ADVANCE program depends upon building networks that connect individual faculty to each other and to administrators. ADVANCE built networks and connected them to existing infrastructure, including academic hierarchy and the WST Center network, but this interconnected system of networks is only effective if individuals participate actively and continuously beyond the boundaries of specific events or activities. Within an environment that incorporates networks and encourages the formations of new ones, individuals must also exercise responsibility to nurture networks for success. An important lesson is that universities interested in institutional transformation must also take care to ameliorate difficult conditions for women while enhancing the environment for all. Any program directed at improving the environment, whether it be directed at women or another minority, must take into account the program's prospects to improve conditions for all faculty.

NOTES

1. "Special Report: The Conundrum of the Glass Ceiling," *Economist*, July 23, 2005.

2. Partners are Iowa State University, Louisiana State University, University of Central Florida, University of Connecticut, Syracuse University, and the University of Utah with collaborators from the University of Texas, El Paso, University of Maryland, University of Pennsylvania, University of Connecticut, University of California, Davis, University of Guelph, Canada, and Kettering University. See www.nsf.gov for a list of ADVANCE awards including the separate grants made to these institutions.

3. The Georgia Tech Center for the Study of Women, Science, and Technology was developed by the colleges of Engineering and Science and the Ivan Allen College, and now reports to the Office of the Provost.

4. A search on Amazon.com results in thousands of entries for "benefits of networking" covering various fields—real estate, computing, community activism, marketing—and stages of work—job hunting, enhancing job knowledge, etc.

5. Mary Frank Fox, "Georgia Tech ADVANCE Survey of Faculty Perceptions, Needs, and Experiences," summary at www.advance.gatech.edu/reports.

6. Jane Ammons is associate dean in the College of Engineering and Mei-Yin Chou is chair of physics in College of Sciences.

7. The ADVANCE team consists of Provost Jean-Lou Chameau (PI); co-PIs Mary Frank Fox, Mary Lynn Realff (former program director), and Sue Rosser; the ADVANCE professors (Fox, Mei-Yin Chou, Jane Ammons, and Mary Jean Harrold); Mary Hallisey Hunt, liaison with provost's office; Carol Colatrella, magazine editor and ADEPT coordinator and current ADVANCE program director; and Angela Shartar, program coordinator.

8. In addition to the colleges including NSF-related disciplines that received professorships under the aegis of ADVANCE, Georgia Tech's administration hopes to provide similar opportunities within its colleges of Management and Architecture in institutionalizing ADVANCE initiatives.

9. See http://www.advance.gatech.edu.

10. Outcomes for twelve NSF-determined indicators:

1. Number and % of women faculty in science/engineering by department
2. Number of women in tenure-line positions by department
3. Tenure promotion outcomes by gender
4. Years in rank by gender
5. a. Time at institution and b. attrition by gender
6. Number of women in S & E who are in non-tenure-track positions (teaching and research)
7. Number and % of women scientists and engineers in administrative positions
8. Number of women S & E faculty in endowed/named chairs
9. Number and % of women S & E faculty on promotion and tenure committees

10. Salary of S & E faculty by gender (controlling for department, rank, years in rank)
11. Space allocation of S & E faculty by gender (with additional controls such as dept., etc.)
12. Start-up packages of newly hired S & E faculty by gender (with additional controls such as field/department, rank, etc.)

REFERENCES

Beaver, Donald B. 2004. Does collaborative research have higher epistemic authority? *Scientometrics* 60:399–408.

Bradley, Raymond T. 1982. Ethical problems in team research. *American Sociologist* 17:87–02.

Fox, Mary Frank. 1991. Gender, environmental milieu, and productivity in science. In *The outer circle: Women in the scientific community,* ed. H. Zuckerman, J. Cole, and J. Bruer, 188–204. New York: W. W. Norton.

———. 1996. Women, academia, and careers in science and engineering. In *The equity equation: Fostering the advancement of women in the sciences, mathematics, and engineering,* ed. C. S. Davis, A. Ginorio, C. Hollenshead, B. Lazarus, and P. Rayman, 265–89. San Francisco: Jossey-Bass.

———. 2000. Organizational environments and doctoral degrees awarded to women in science and engineering departments. *Women's Studies Quarterly* 28:47–61.

———. 2001. Women, science, and academia: Graduate education and careers. *Gender and Society* 15:654–66.

———. 2003. Gender, faculty, and doctoral education. In *Equal rites, unequal outcomes: Women in American research universities,* ed. L. Hornig, 91–109. New York: Plenum/Kluwer Academic.

Fox, Mary Frank, and Carol Colatrella. 2006. Participation, performance, and advancement of women in academic science and engineering: What is at issue and why. *Journal of Technology Transfer* 13:377–86.

Hornig, Lilli S., ed. 2003. *Equal rites, unequal outcomes: Women in American research universities.* New York: Kluwer.

Jamieson, Kathleen Hall. 1995. *Beyond the double bind: Women and leadership.* New York: Oxford University Press.

Kanter, Rosabeth Moss. 1977. *Men and women of the corporation.* New York: Basic Books.

Moody, Joann. 2004. *Faculty diversity: Problems and solutions.* New York: Routledge.

Pelz, Donald, and Frank Andrews. 1976. *Scientists in organizations: Productive climates for research and development.* Ann Arbor, MI: Institute for Social Research.

Presser, Stanley. 1980. Collaboration and the quality of research. *Social Studies of Science* 10:95–101.

Rosser, Sue V. 2004. *The science glass ceiling: Academic women scientists and the struggle to succeed.* New York: Routledge.

Valian, Virginia. 1999. *Why so slow? The advancement of women.* Cambridge: MIT Press.

Wray, K. Brad. 2002. The epistemic significance of collaborative research. *Philosophy of Science* 69:150–68.

SIX

A Faculty Mentoring Program for Women

BUILDING COLLECTIVE RESPONSIBILITY FOR A
HIGHLY QUALIFIED FACULTY

Evelyn Posey, Christine Reimers, & Kelly Andronicos

A FEMALE UNTENURED faculty member at the University of Texas at El Paso (UTEP) once wrote, "I have been consistently left off of lists of new faculty candidates for service opportunities in the college and university. My bosses say that 'I am out of luck.' What message does that send?" Another wrote, "Already I am tired of being given conflicting advice on what is required for tenure. I am afraid that in a few years I just won't care." Comments such as these, a review of UTEP data that underscored the need to recruit and retain more women faculty, and research that demonstrated the success of mentoring programs at other universities served as the impetus for a UTEP Faculty Mentoring Program for Women (FMPW). Our motivation for starting such a program was to improve the working environment for women faculty, which would in turn improve the working environment for all. We were committed to the belief that a diverse faculty is a qualified faculty and that women bring new research talents and new ways of doing things to traditional college departments.

Begun in fall 2000, the FMPW was designed to support the professional advancement of newly hired tenure-track women in all academic departments, Science, Engineering, Technology, and Mathematics (STEM) and non-STEM, and to increase institutional capacity for maintaining a highly qualified faculty. Now part of the National Science

Foundation (NSF) ADVANCE Institutional Transformation for Faculty Diversity initiative, the program is designed to meet three goals: (1) increase women faculty's effectiveness and visibility through improved access, (2) facilitate the attainment of individual career goals, and (3) orient women faculty to the culture and organization of the institution. A small group of faculty on the Women's Advisory Board to the President wrote the original proposal that launched the program in fall 2000. The stated mission: "to assist women faculty in their professional development through the guidance and support of experienced UTEP faculty members who serve as role models, advisors, and advocates."

The authors of the proposal envisioned a program that would benefit the institution through increased hiring, retention, and promotion of women faculty; increased productivity and job satisfaction; a climate of collegiality and cooperation; and the opportunity to nurture future institutional leaders. Benefits to mentees would include assistance in defining career goals that could improve chances for tenure and promotion, improved access to UTEP information and resources, a better understanding of the UTEP organizational structure and culture, and practical advice on balancing the faculty workload with the sometimes conflicting demands of family. Benefits to mentors would include the satisfaction of having helped a colleague, contributed to the overall success of the university, increased the potential for joint research, and improved mentoring skills.

Continued Need for the Program

Research shows that while higher education is doing a decent job of graduating women with research doctorates, the top research institutions are still not doing a very good job of hiring them. The few women who are hired advance more slowly than men, are paid less, and are more dissatisfied with their jobs (Wilson 2004). Studies on faculty retention and advancement have suggested that women's experiences and perceptions of academe are quite different from those of their male colleagues (Johnsrud and Atwater 1993; Johnsrud and Des Jarlais 1994; Harper et al. 2001). For women, the decision to leave or stay at an academic institution most often hinges on their satisfaction with working conditions. Women mention that workplace satisfaction is affected by work conditions, including negative factors such as being actively discouraged from participating in departmental and institutional decision-making (Trower and Chait 2002); having to succeed by playing by even more stringent rules than their white male colleagues; lack of attitudinal support for their research (Wenzel and Hollenshead 1994; Johnsrud and Atwater 1993; Honeyman and Summers 1994); noncollegial and non-collaborative climates within the departments; greater social isolation; difficulty balancing teaching with other responsibilities (Johnsrud and Atwater 1993; Olsen and Maple 1993; Ruffins 1997); lack of guidance on tenure expectations and realities; alienation from the institution; lack of opportunities in research; lack of respect from colleagues; lack of support from peers and the administration (Trower and Chait 2002; Wenzel and Hollenshead 1994; Johnsrud and Atwater 1993; Borman et al. 1998); and dual career situations where the spouse did not get assistance (Wenzel and Hollenshead 1994).

The realities of these women's lives lead to a feeling of marginalization (CAWMSET 2000; Borman et al. 1998), which grows greater as women progress in their careers past tenure. While junior women initially believe they are treated equally, senior women point to subtle patterns of discrimination and neglect that affect satisfaction with working conditions at their institutions (CAWMSET 2000; Baker 2003). And, when there are efforts to address these inequities, to increase numbers of women faculty, and to improve departmental climate and diversity, they are confounded by charges that such efforts smack of affirmative action and lessen the quality of the faculty (Jackson 2004).

Since the divisive comments made by Harvard university president Lawrence H. Summers, who suggested that there were innate differences between men's and women's abilities in mathematics and science, the merit of such mentoring programs has been debated not only in university halls, but in the national news as well. Partly in response to the uproar surrounding his comments, Summers appointed two panels: the Task Force on Women Faculty and the Task Force on Women in Science and Engineering to study what might be done to recruit and advance women faculty. The latter committee, chaired by Barbara J. Grosz, has as one of its recommendations to "improve mentor and advising programs for postdoctoral fellows and junior faculty members" (Fogg 2005).

Original UTEP FMPW Model

The key elements of the UTEP program were modeled after the University of Wisconsin–Madison's Women Faculty Mentoring Program, started by Molly Carnes and Lindsey Stoddard Cameron, whom we invited to the UTEP campus during the first year of our program. In preparation for writing the proposal, we had conducted an exhaustive literature review; it was through this review that we learned of the Wisconsin program. At the time, we had no idea that both institutions would eventually become part of NSF ADVANCE. Indeed, most of the NSF ADVANCE institutions include faculty mentoring, and many have special programs designed specifically for this purpose.

The UTEP FMPW pilot, managed through the Center for Effective Teaching and Learning under the direction of Dr. Christine Reimers, was introduced as part of the new faculty orientation. All women, both STEM and non-STEM faculty, hired in tenure-track appointments for the fall 2000–2001 academic year participated in the program. Mentors, who could be men or women, were recruited from the colleges in which new tenure-track women were hired. Each potential mentor or mentee completed an interest survey and each mentee was then assigned two mentors: one from within her college and one from another college, one male and one female. The one from inside the college would be able to respond to questions about research and college realities, while the one from outside the college would be a safe person to talk with about other subjects that the mentee might not want aired in her own college or department. Providing one male and one female mentor was intended to help us see whether gender had an effect on the quality of the mentoring relationships.

In addition to their own formal and informal meetings throughout the year, mentors and mentees participated in monthly events—seminars, discussions, and luncheons—that were designed to help mentoring pairs communicate effectively and to further their discussions of access, balance, and acculturation to UTEP. Topics for such sessions included, among others: communication strategies, professional and personal balance, discussions on cutting through red tape to achieve results, and dealing effectively with administrators.

By the end of the pilot year, it was clear that assigning two mentors, one from the mentee's college and one from another college, was only partially effective. Mentees reported difficulty in establishing meaningful relationships with faculty mentors outside their colleges. Lack of proximity played a significant role, as did lack of compatible disciplinary interests. One mentor from another college, for example, stated, "Even though I have explained research requirements, she does not seem to understand what research is." Later, this mentor acknowledged that she just didn't understand the nature of the research conducted in the mentee's discipline.

These two factors, proximity and disciplinary compatibility, were key in determining the quality of the mentor–mentee relationship and proved to be more critical than did the gender of the mentors. There was no information to suggest that female mentors provided an inherently better mentoring experience than did their male counterparts. Because of these revelations, the mentoring model was changed. Each new woman faculty member was now assigned one mentor, male or female, from within her college but outside her department. Mentors have always been chosen outside the department so that any mentoring activities that might already be taking place within departments not be displaced, and so that the mentees have someone from whom to seek advice who is not influenced by departmental politics.

Program assessment in the spring of 2002 revealed a perceptual mismatch in a significant number of mentor-mentee pairs that resulted in critical communication gaps between partnered individuals. In some cases, mentors reported their mentee's progress in the first year as positive and were reluctant to seem "pushy" by "interfering" in the mentee's life. One mentor described it this way: "I've made it clear that I'm available at any time for consultation, but my mentee has not taken me up on that offer." Conversely, mentees reported that more contact would be useful but that they, too, were reluctant to initiate the meetings because of the perception that mentors were too busy. In certain cases, this concern for the mentoring partner's freedom, time, and independence lessened the impact and usefulness of the mentoring relationship.

As a result, more explicit instructions regarding making regular contact were developed for both mentees and mentors, pressing both parties to become more proactive in this area. A new goal-setting workshop was developed for mentees that they attended before they met their mentors. This session was designed to help mentees gain better insight into their current professional standing, assess their satisfaction with the balance between their personal and professional lives, set specific goals for the first several years of their appointment at UTEP, and outline expectations for their mentoring relationship. Additionally, a mentor orientation was developed. Initially, we believed that mentors needed only minimal training in how to establish and maintain a mentoring relationship. Since mentors were already highly successful in their disciplines, there was a reluctance to treat them as if they, too, needed mentoring. Mentors, however, while successful in the professoriat, were not innately good mentors. In order to address this deficiency, an in-depth mentor orientation was designed to provide guidance and clarification of the mentor's role. With a few years of program experience to draw upon, we were able to point to trends that had developed in the mentees' behavior. For example, in addition to mentees' reluctance to "bother" their mentors, there was a trend for mentees to be very involved in the program for the first year, but to become much less involved in the second year due to increasing time constraints. This pointed to the need for those mentors who intended to continue their relationships past the first year to provide continued encouragement and to persist in communicating with their mentees at a time when the new women faculty might feel particularly overwhelmed by the demands of research, teaching, and university service.

In an additional effort to bolster the communication flow, a January workshop was added to provide another opportunity for mentoring pairs to connect (if the original connection in the fall was not successful) or to reconnect for those pairs whose relationships might have stalled. Mentoring goals and previously agreed upon meeting schedules were revisited and discussed.

This same assessment yielded information about the importance of peer mentoring developed through interactions at organized FMPW events. Mentees indicated that a structure that accommodated more informal contact among mentees from all years would be welcome. On the year-end evaluation, one mentee wrote, "Have events for new faculty that allow us to speak to each other and that provide more opportunities for social interaction." In response to comments such as these, monthly brown bags were added to the roster of formal workshops and events in fall 2003. Participation was limited to current and past mentees and the program director. Often a topic based on an article or an issue of interest, such as information on grant writing or promotion and tenure, was prepared beforehand. When no topic was designated, mentees initiated discussions on subjects such as managing authorships on papers, finding interdisciplinary collaborators, and dealing with difficult colleagues or situations in their home departments.

It also became apparent that using an interest survey to assist in making successful matches often led to unreasonable expectations. The survey solicited information regarding professional interests and expertise plus information about personal interests such as culture, religion, and family issues. This led some mentees to believe that a mentor match could be made taking into account all these various personal factors. In reality, due to the perennial shortage of qualified and willing mentors, the mentees often had the potential to be paired with only a handful of mentors from their colleges. Therefore, personal considerations usually could not play a significant role in the matching process. Consequently, the use of these types of forms was abandoned.

Up until this point, the participants agreed to remain active in the program and maintain formal contact with their partners for the duration of the academic year. As partnerships were not formed until September and because the month-long winter break often disrupted the continuity of the new relationship, participants reported that the time allocated was insufficient to reap substantial mentoring benefits. As a result, the twelve-month model was expanded to eighteen months.

Current Program Model

Given the increase in the number of new women tenure-track faculty hired, partly as a result of the successful ADVANCE initiative begun in fall 2003, the difficulty of finding qualified mentors increased each year. By fall 2004, the FMPW had matched eighty mentees with faculty mentors. Some mentors had already served twice. After four years of making one-on-one mentor/mentee matches, now expanded to eighteen months, circumstances dictated a radical change from the original model. Because mentors must be nominated by peers or chairs and then vetted for their success in teaching, research, and service and their likely aptitude for mentoring, the program began to face a significant shortage of mentors. Because the University of Wisconsin's program has been in place since 1989, they have a number of women who participated in their mentoring program who are now serving as mentors. As UTEP women who have participated in the FMPW are tenured and promoted, it is our hope that they too will give back to the program in the future. Meanwhile, in an effort to address the shortage of mentors, the program adopted a group-mentoring model. The group model capitalizes on the organic peer-mentoring phenomenon while managing an insufficient number of mentors. Generally, up to six mentees from the same college and sometimes from the same department are matched, as a group, with two mentors from their college, but from outside the mentees' departments. This new model, still requiring an eighteen-month commitment, seems to have relieved some of the anxiety over "marrying strangers" that one-on-one partnering agreements sometimes produced. Mentees have more opportunity to exchange views with peers, and mentors have a partner-mentor with whom they can discuss sensitive situations or questions that arise with their mentees.

The group-mentoring model replaces formal monthly workshops with more informal monthly group luncheons, where the entire mentoring group assembles to discuss matters of interest. In some cases, mentees have been grouped across colleges when the disciplines are similar enough to be congruous (Psychology, Sociology/Anthropology, and Educational Psychology and Special Services, for example). However, the model includes flexibility to accommodate women from departments where they are traditionally underrepresented and where sometimes there is only one woman hired—a phenomenon found only in STEM departments. In these circumstances, one-on-one matches may still be arranged. In addition, monthly brown-bags continue to be an efficient venue to deliver information on topics such as how to negotiate the challenges of maternity and progress toward tenure, definitions of leadership, and development of teaching portfolios. September and October are still reserved for formal sessions that cover essential program information. These include

- Mentee/Mentee meeting where new women faculty are introduced to the program and network with existing program participants
- Mentee goal-setting session where mentees address short- and long-term goals, as well as their own expectations for their mentoring relationships
- Mentor orientation that provides guidance and a program overview for those faculty serving as mentors.

While the group-mentoring model is only a year old, a preliminary evaluation reveals some of the strengths of the new model. In talking about the benefits and support provided by the program, one mentee stated, "My mentoring team has been great, providing both social and professional support." Mentors, too, indicate that they benefit from participating in the program. One states, "It was fun and not a lot of work. I gained more or as much from the mentees [as] they got from the program. It was inspiring to work with new, enthusiastic scholars."

Challenges

The challenges for the UTEP Faculty Mentoring Program are fourfold. They include (1) the difficulty of coming to an institutional definition of success for the mentoring program; (2) addressing potentially negative perceptions of a mentoring program, and in particular a mentoring program for women; (3) barriers related to maintenance and expansion of campus-wide mentoring programs; and (4) effective program assessment that leads to institutionalization of such programs.

Definitions of Success

As described above, the Faculty Mentoring Program for Women at UTEP was instituted because an active group of faculty women recognized the need for mentoring for pretenure women. The reasons for such a need are various, but a major assumption was that the university may be losing qualified women. Unfortunately, this underlying assumption leads to tension between differing interpretations of what "success" of the mentoring program means, and thus how to measure it. Clearly, the easiest measure of success would be a higher retention rate of women through tenure and promotion. Thus, at first glance, it would seem to be sufficient to count the numbers of women hired, the numbers of women who leave before tenure and the numbers of women accorded and denied tenure, and then compare these numbers to previous cohorts of women who did not have the benefit of the mentoring program. There are several problems with this approach, however, which did not become clear until several years into the program. First of all, the university does not document the loss of women (or men) before tenure. That is, faculty who were hired as tenure track but who left the university before applying for tenure are almost impossible to trace. Because the university keeps no record of their exit, there is often no way to know that they were there in the first place. Nor is there a coherent way to find out why these faculty left. This means that, without intensive detective work, we have no way of comparing past cohorts with those who have benefited from the FMPW.

Additionally, the reasons for leaving an institution are multiple and varied: many do not fall within the purview of any mentoring program. Mentoring programs can only help to acculturate faculty to the institution, provide advice on how to document and communicate successes, and give ideas on how best to organize a professional career.

We have found, however, that the reasons for leaving the university range from the lack of professional possibilities in the region for a spouse, to a more prestigious institution able to offer higher salaries, to a chilly climate in the department, to the realization that academe (or the institution) is not a good match for the candidate's aspirations in life. Thus even if we were to find accurate retention rates for pre-FMPW and post-FMPW cohorts, the results may not reflect the success of the program.

Pure retention and advancement rates do not measure the success of the program for other reasons as well. Some of the women in the program credit their success and retention in part to the support provided by the program. Yet are we to say that a woman who decides that a different institution is more likely to provide a productive professional home than her current one, but who decides it after one year instead of after five years spent unproductively, has not been positively influenced by the program? This has been the case with some faculty women. In one case, for example, a woman faculty member was unhappy with her teaching load. She was much more at ease doing research than providing instruction in the classroom or mentoring students. At the end of her first year, and as a result of conversations with her mentor and with peers in the FMPW, she decided to go to a different institution where she could concentrate more fully on her research. Two years later, she contacted the director of the FMPW in order to learn more about the program so that she could institute one at her new university. Clearly, the FMPW had a seminal role in the evolution of this woman's academic career. But when only retention data are valued as indicators for success, this woman counts as a defeat for the program.

Addressing Negative Perceptions

Mentoring programs invite questions about their intentions and their value, especially when the success of such a program is hard to define. Countering negative perceptions of the program can be a difficult task, especially in a context where the idea of "affirmative action" has become a politically charged one, and where a mentoring program focused on women may be seen as catering to a minority at the expense of academic rigor and quality. While the program has undeniably provided benefits to participants, questions about the program that we have heard include the following:

1. "Why do women need fixing? Isn't this just another way to isolate us?"

2. "Why is this not a mentoring program for men, too? Don't we all need mentoring?"

3. "Why should women participate in this program—if they are already as stretched for time as you say, isn't this just another way for them to lose time from their research?"

4. "Shouldn't professional adults be able to stand on their own two feet—are you not enabling them?"

Clearly, the first objection to the program is one of bias avoidance. Women, already under pressure in their discipline for being women, may want to avoid any appearance of special treatment that could incite their colleagues' jealousies or accusations that they are not as fully qualified as their male counterparts but are only getting by with extra help. Often, younger women just beginning their careers are particularly prone to this view. They start out believing that equal treatment has already been achieved, and due to lack of experience they do not yet see institutionalized bias working against them in their careers. It is clear from the Massachusetts Institute of Technology report (1999) that it is only when women have persevered over years and compare their experiences later in their careers that patterns of discrimination become clear to them. Since the FMPW is aimed at women in their first years at the university, this perception is difficult to overcome. To counter it, it is useful for new women faculty to hear from experienced women in academe so that they can see the implications of patterns they see developing in their home departments. Of course, when there are few experienced women to talk to, this becomes an even more difficult task.

The second objection is, in one sense, reasonable. Of course, all faculty members new to an institution need mentoring. In fact, the UTEP ADVANCE external advisory board has strongly recommended that our program be extended to men, precisely to avoid a backlash against a program for a selected group, and to extend its documented benefits to all faculty members. However, this objection does not take into account the lack of traditional mentoring networks available to women. It is of course difficult to see a lack of a network, especially when one has such a network oneself—young men may be unconsciously taking advantage of their mentoring networks and be entirely unaware that such options are not available to women to the same degree. One way to counter this objection is to expand the program to men, but to retain some aspects of the program for women only. This has the obvious disadvantage of still looking preferential.

The third and fourth objections stem from a basic misunderstanding of the power of mentoring. Unfortunately, there are many faculty members and administrators in academe that hold these views. It is true that getting advice, talking with peers and with experienced colleagues, organizing one's professional career by identifying goals and realistic time-lines initially takes extra time. But time spent in this way early on is an investment in the shape of a whole career, and thus in professional success over the long term. A faculty member who can navigate the culture and organization of an institution while keeping his or her own goals in mind is clearly at an advantage over those who do not have a career strategy in place. The idea that faculty members should be able to stand on their own two feet and that mentoring enables the weak ones replicates a kind of hazing that we deplore in other types of organizations. Mentors in the program regularly report that, even as successful and experienced faculty members, they still need mentoring in some areas of their professional lives. This is so because a faculty life has, in effect, seasons. The first year of a pretenure career does not look like the third year, nor like the first year after promotion, because expectations of faculty members change as they progress in their careers. The priorities of research, teaching, and service quite naturally ebb and flow, depending on where in her career a faculty member is.

Maintenance and Expansion of the Mentoring Program

Ideally, the FMPW should be expanded to include all pretenure faculty members. However, there are several barriers that any mentoring program will face when envisioning such an expansion. First, any mentoring program may eventually face a shortage of qualified mentors for the number of new faculty the institution is hiring. Second, if the program is both popular and centralized, as is the case with FMPW, which is housed in the Center for Effective Teaching and Learning, the program will face a lack of infrastructure that will make it very hard to maintain current efforts, much less expand.

The first challenge of lack of qualified mentors is a serious one. Mentors must be good teachers, have an active research program, and possess the personal qualities required for good mentorship of a junior colleague. Faculty members who are recognized for these attributes are regularly asked to join many other projects at the university. They have both the personal and organizational skills that allow them to become excellent chairs of committees, chairs of departments, and heads of task forces. This means that the pool of mentors available becomes quite restricted because people with these characteristics are already over-loaded with service in other areas of the university, or are already serving as mentors in the program. Mentors are expected to remain in their formal mentoring relationships for eighteen months. Many have sponta-neously continued their relationships all the way through the mentee's tenure decision, thus making it more difficult for them to add another mentee to their load. An added disincentive for mentors is the lack of rewards for mentoring. Currently, in addition to the intrinsic satisfaction they get from their relationship and from the opportunity to help a junior colleague, mentors receive the gift of a book and a pen.

In order to achieve a higher profile and reward system for mentor-ing, we are currently working with the deans of the colleges to include it as an explicit dimension of service in the annual merit reviews. When it becomes a part of the expectations of faculty and when it is attached to the reward system in place at the university, we may find more senior faculty members willing to spend their time in this manner.

The lack of infrastructure for mentoring is a serious barrier. With over seventy participants (mentees and mentors) in the current program, it has grown beyond the capacity of one person to direct it. In response, ADVANCE has established ADVANCE faculty fellows in each college, senior faculty members who have been given small stipends and addi-tional resources to help with the recruitment of mentors and coordina-tion of the semimonthly events. The current director will continue to oversee the entire program and work closely with these faculty fellows.

Program Assessment for Institutionalization

In order to make appropriate arguments to institutionalize the program (i.e., giving it a permanent source of funding) the program needs a thor-oughgoing assessment. In the past, the program has paid an external evaluator a stipend to prepare annual reviews of the program, for both formative and summative purposes. These reports have been useful to help make arguments each year for the program's small budget to be renewed, as well as to provide information for needed changes to the structure of the program. However, after five years of the program's existence, it is time for a more comprehensive look at its benefits and challenges if it is to remain viable and be incorporated into the university structure.

Such a rigorous assessment of the program would require (in addition to the currently administered questionnaires and focus groups of participants who are participating in the program) a comparison of pre-FMPW cohorts with the current one, in terms of both retention rates and perceptions of support as they began their careers; a comparison of FMPW participants with their male counterparts along those same lines; and contact with every individual who has been a participant in the program to measure perceptions of its influence on her development, whether she was retained at UTEP or not. Such a meticulous review of the program will require significant time from an evaluator, for which there must either be a budgetary allocation or a time allocation for the evaluator of the ADVANCE initiative, who already has many projects to satisfy both NSF's and our own requirements for demonstrating success.

Conclusion and Recommendations

Through program assessment, we have striven for continuous quality improvement and accountability, and although our successes have been great, we continue to revamp the program to maximize its benefits. Based on what we have learned, the following are some recommendations for establishing such faculty mentoring programs.

Find an upper-level administrator who will support the program verbally and financially. Grassroots efforts are valuable, but without such administrative support, the program cannot be sustained.

In our experience, simply supporting new faculty by providing information and resources to them, responding to their questions and needs, and helping them develop networks with their peers and senior faculty—while of long-term benefit to individual faculty members—is not enough for long-term success of a mentoring program. Such a program also needs the explicit support—both verbal and financial—of upper administration, to make institutional change a reality. Notwithstanding the very positive reports of the participants and their chairs, some administrators still believe that a mentoring program is an "extra" that might "enable" weaker faculty members and lure them into false hopes for success at tenure time. Additionally, if "word of mouth" news of successes does not reach administrators and change language and attitudes, there is little hope either for future financial support or for directing the institution toward what one of our deans has hopefully called a "culture of helpfulness."

Ensure proper rewards for mentoring within the university reward system.

At our institution, senior faculty members are stretched to capacity with the demands of teaching, research, and service. As is the case in most academic institutions, mentoring is one of those critical activities that "just happen" according to the informal behaviors of the department in question. Senior faculty members are not explicitly rewarded for putting in their valuable time to help orient their younger colleagues. They are explicitly rewarded for their efforts to secure grants, do research, excel in teaching, and, to some extent, serve on committees.

Find ways to incorporate existing departmental and college structures, so that women faculty aren't sent "over there" for mentoring.

This point is a direct corollary of the previous one. New faculty quite rightly see the department as their home and the most direct source of information on how to succeed. This is the function of a department, and in particular of the department chair. However, when mentoring has not made it into the culture of a department (or when the chair has been selected for his or her excellence in research, but not for personal, management, and communication skills), new faculty members will struggle, not understanding the proper directions to take in order to become effective and advance in their profession. The "culture of helpfulness" needs to permeate where people live, that is, the departments and the colleges. UTEP's Center for Effective Teaching and Learning is well respected by the faculty and administrators, but when someone outside the department structure can provide mentoring, the temptation is strong to let it happen elsewhere, and conserve the time and energy it would take to correctly orient and mentor the department's own faculty. Responsibility for effective mentoring should lie with those people who will be working most closely with the new faculty.

Ensure that there is a mentoring coordinator or director to ensure consistency and quality programming across the colleges.

Because there is not a tradition of mentoring that is equal across all departments and colleges, it is advisable to retain a central coordinator or director. Such a person must ensure that high-quality programming and advice is being provided to all faculty, and that faculty members in depart-ments where the culture of helpfulness has not taken root have access to appropriate information and resources despite their local situation.

Include peer mentoring opportunities, not just one-on-one mentoring.

After five years of our mentoring program, it is very clear that the networking and peer mentoring that happen in the mentoring teams and during the brown-bags is at least as important as the direct mentor-ing and advice new faculty receive from their assigned mentors. New faculty find it important to know that they are not alone in facing chal-lenging demands, and they are remarkably creative in the solutions they provide to one another when they are given the opportunity to brain-storm and troubleshoot in a safe environment. Additionally, when a program pairs or groups strangers, there is the inevitable danger that the mentor-mentee relationship may not gel, either because of limitations on the time available for the relationship on the part of the partners, or because the mentoring partners discover that they don't really work well together.

After five years of experience with UTEP's Faculty Mentoring Pro-gram for Women, we are confident that we have helped strengthen individual and institutional support for women and found new ways to maintain a highly qualified and diverse faculty. As a result of the pro-gram, fewer and fewer women express concern about what is expected of them for tenure and promotion or how to balance research, teaching, and service. Instead, we hear comments about the value of the program and the university that offers it. One woman describes her decision to come to UTEP as a result of hearing about the FMPW from a colleague: "She spoke to me about the potential to achieve at UTEP; tempted me with the possibility of being able to collaborate on various international projects with her; and mentioned her own experience with the strong support new faculty receive at UTEP, particularly faculty women. She made me realize if a university goes to such great lengths to support women faculty, it must be a positive place to grow."

REFERENCES

Baker, Christina L. 2003. *What do women want? Equity.* New England Board of Education: EBSCO Publishing.
Borman, K., J. D. Kromrey, D. Thomas, and W. Dickinson. 1998. University women and minorities: A case study of organizational supports and impedi-

ments for faculty. Paper presented to the American Educational Research Association, San Diego.

Commission on the Advancement of Women and Minorities in Science, Engineering, and Technology Development (CAWMSET). 2000. Land of plenty. http://www.nsf.gov/od/cawmset/start.htm.

Fogg, Piper. 2005. Harvard creates 2 panels to advance female professors. *Chronicle of Higher Education* 51 (24): A12.

Harper, E. P., R. G. Baldwin, B. G. Gansneder, and J. L. Chronister. 2001. Full time women faculty off the tenure track: Profile and practice. *Review of Higher Education* 24 (3): 237–57.

Honeyman, D. S., and S. R. Summers. 1994. Faculty turnover: An analysis by rank, gender, ethnicity and reason. Paper presented at the Eighteenth National Conference on Successful College Teaching, Orlando, FL.

Jackson, J. 2004. The story is not in the numbers: Academic socialization and diversifying the faculty. *NWSA Journal* 16 (1): 172–84.

Johnsrud, L. K., and C. D. Atwater. 1993. Scaffolding the ivory tower: Building supports for new faculty to the academy. *CUPA Journal* 44 (1): 1–14.

Johnsrud, L. K, and C. D. Des Jarlais. 1994. Barriers to tenure for women and minorities. *Review of Higher Education* 17 (4): 335–53.

Massachusetts Institute of Technology. 1999. A study on the status of women faculty in science at MIT. *MIT Faculty Newsletter* 9 (4). http://web/mit/edu/fn1/women//women.html.

Olsen, D., and S. A. Maple. 1993. Gender differences among faculty at a research university: Myths and realities. *Initiatives* 55 (4): 33–42.

Ruffins, P. 1997. The fall of the house of tenure. *Black Issues in Higher Education* 14 (7): 19–26.

Trower, C. A., and R. P. Chait. 2002. Faculty diversity: Too little for too long. *Harvard Magazine,* March–April. http://www.harvard-magazine.com.

Wenzel, A., and C. Hollenshead. 1994. Tenured women faculty: Reasons for leaving one research university. Paper presented at the Annual Meeting of the Association for the Study of Higher Education, Tucson, AZ.

Wilson, R. 2004. Where the elite teach, it's still a man's world. *Chronicle of Higher Education* 51 (15): A8–12.

SEVEN

Beyond Mentoring

A SPONSORSHIP PROGRAM TO IMPROVE WOMEN'S SUCCESS

Vita C. Rabinowitz & Virginia Valian

WOMEN LAG BEHIND men in all the professions, including academia and science. In academia, women progress through the ranks more slowly, are tenured more slowly, and make less money (American Association of University Professors 2005; Long 2001; National Science Foundation 2004; Valian 1998); those problems are exacerbated at research-intensive institutions. Relative to men, women scientists also publish somewhat less, obtain less information about how to succeed, receive less support for their careers, and get less recognition for their accomplishments (see Valian 1998 and references therein); those problems are exacerbated at the teaching-intensive institutions where women are overrepresented (for the most recent data, see Cataldi, Bradburn, and Fahimi 2005).

At teaching-intensive institutions, both male and female researchers are unlikely to be as productive as their peers at research-intensive institutions: they have high teaching and advising responsibilities, few teaching assistants, few or no graduate students, and, often, substandard research infrastructures. At the City University of New York, for example, of which Hunter College is a part, the contractual instructional workload for full-time, tenure-track faculty in Arts and Sciences is seven courses per year. Nationally, full-time faculty at private baccalaureate institutions—of whom 41% are female—spend 68% of their time teaching and 12% of their time on research; their peers at public master's institutions—of whom 41% are also female—spend 66% of their time teaching and 14% of their time on research; those at doctoral institutions—of whom only 32% are female—spend 50% of their time teaching and 28% of their time on research (Cataldi, Bradburn, and Fahimi 2005). Although limitations at teaching-intensive schools confront all faculty, gender and minority status intensifies those problems: women and people of color are likely to receive even less information, fewer resources, and lower evaluations than their white male peers.

The Gender Equity Project (GEP) at Hunter College, partially funded by the National Science Foundation's ADVANCE program, addresses the challenges women face in academia, particularly at teaching-intensive institutions where women and minority scientists are overrepresented and where female, first-generation-college, and underserved minority students are most likely to be educated. The Sponsorship Program was designed for the faculty at Hunter College, but most of its components will work well at any institution.

In this chapter we describe the current form of the Sponsorship Program and how it has developed. The program at present offers a combination of (a) financial support, (b) paid sponsors, (c) workshops, and (d) ongoing consultations with GEP codirectors. Although this combination is especially attractive, the components can be used separately, depending on the features of the institution.

The Sponsorship Program provides participants with time and resources for research, as well as opportunities to (a) pursue new research areas, (b) acquire new research or technical skills, (c) begin different types of scholarship, (d) interact with women scientists from a variety of fields, (e) learn more about developing and advancing their careers, and (f) discover and share successful strategies for academic success.

We present evidence that the program has improved women's productivity, advancement, and identities as researchers, and we discuss ongoing challenges. We conclude with six principles that have emerged from our work in the GEP in general and the Sponsorship Program in particular.

The Gender Equity Project at Hunter College

The GEP operationally defines a scientist as anyone whose research would be eligible for NSF funding from a research-based directorate. We thus include in our purview five natural science departments and six social science departments within Arts and Sciences. The five departments in natural sciences are Biological Sciences, Chemistry, Computer

Science, Mathematics and Statistics, and Physics. The six departments in social sciences are Anthropology, Economics, Geography, Political Science, Psychology, and Sociology.

By many measures, women fare well at Hunter compared to men: they are well represented in the upper administration, among department chairs, and among full professors (in 2004–5, women were 23% of full professors in natural sciences and 27% in social sciences); they are 83% of the Distinguished Professors in the sciences (a rank to which fewer than 2% of CUNY faculty are appointed); their salaries are equivalent to men's. At the same time, compared to men, women leave more often at the assistant professor level, receive less start-up funding, and spend more time at the rank of associate professor. Informal observations suggest that all three problems are common across US colleges and universities.

The GEP's Sponsorship Program

The main goal of the Sponsorship Program is to enhance the research productivity and academic stature of women engaged in basic science at Hunter College. The program is open to Hunter women scientists of any rank and at any point in their careers. From June 2002 through May 2006, twenty-six women (38% of all women scientists at Hunter), representing all academic ranks and nine of the eleven departments, have participated in the program as associates. Associates can receive support for a total of three years, which need not be contiguous.

The Application Package

Candidates apply year by year for a maximum of three years. The application process achieves three aims: it identifies strengths and weaknesses in an applicant's portfolio, it models professional grant writing and thus provides practice for new investigators, and it establishes the program as desirable and selective rather than remedial. Applications require a curriculum vitae, statements of past, present, and future scholarly interests, resources needed for the coming year, research goals and commitments for the coming year, a budget and budget justification, a statement of other sources of funding, a description of the ideal sponsor(s), and a letter of support from the department chair. After the first year of participation, applications must include a statement of the benefits of prior

Sponsorship Program participation. The GEP codirectors personally interview all applicants.

Because social science research suggests that written commitments help ensure desired behavior (Levy, Yamashita, and Pow 1979; Wurtele, Galanos, and Roberts 1980), the application package also requires participants to sign a contract committing themselves to a set of goals and activities for their sponsorship year. This includes general goals like attending GEP workshops and colloquia, meeting with the codirectors periodically to refine goals and review progress, and consulting with the sponsor on a biweekly basis, as well as specific goals such as submitting specific grants and papers and acquiring new knowledge or techniques. The GEP codirectors evaluate the proposals on the basis of the intellectual quality of proposed research, the extent to which program participation is likely to increase (or has already increased) scholarly productivity, appropriateness and feasibility of proposed goals and commitments, and suitability of the proposed budget.

The diagnostic and educational value of the application package was clear from the start: we were able to evaluate applicants' grant-writing skills and determine whether they needed more training in this activity; applicants gained valuable experience in grant writing. It was clear from the first set of applications that potential associates needed more guidance in grant writing. We thus introduced changes in the application guidelines. In order to model the desired result, the GEP's director of programs and research, Annemarie Nicols-Grinenko, composed a sample application that covered all the complications of a proposal. The quality of applications improved greatly as a result of this model.

Who Is Served

The Sponsorship Program is both a research project aimed at understanding career trajectories and an intervention aimed at improving women's productivity. In year 1 we accepted all twelve women who applied. Of those twelve, eight were assistant professors and four were associate professors; their dates of degree ranged from 1979 to 2000. In years 2–4, a similar proportion were assistant and associate professors. In year 2, a full professor was accepted into the program. Overall, the program has attracted more women from social science than natural science, but we have found that this difference is less important than the laboratory/nonlaboratory division. One workshop, for example, covers managing a lab, which is not directly useful for nonlaboratory scientists.

Financial Support

Associates can receive up to $10,000 per year. They may use those funds for release time if their chair agrees, research assistance, teaching assistance, travel, or equipment—anything that is research-related and approved by the GEP. This portion of the Sponsorship Program is the equivalent of an internal grant to the faculty member. It is one way that an institution can indicate its support and commitment to a faculty member. A few external grants, such as one for $15,000 that the Henry Luce Foundation provides to junior women, serve a similar purpose.

First-year associates typically receive the full $10,000. In order to support as many deserving faculty as possible, partial funding often occurs in years 2 or 3. The funding is the biggest attraction of the program, and, in evaluations of the program, associates rate funding as its most valuable aspect. Although the dollar amount is small, it can be used to purchase time, which is the resource most important to faculty with heavy teaching responsibilities.

We have made it explicit that release time can only be used for research purposes, not for outside teaching or consultation. In a few cases, we have made monetary awards for only one semester, with continuation for the entire year contingent upon making progress in specific areas.

Sponsors

Sponsors serve as intellectual sounding boards, provide critiques of papers and grant proposals, make suggestions about where to submit papers and grant proposals, recommend conferences to attend, and provide strategic advice. Sponsors receive financial support (up to a total of $2,500 per associate per semester) for their participation in the program. A GEP codirector speaks once a semester to each sponsor.

We define a sponsor as a successful senior male or female in the associate's field, but not in her home department. The potential for perceived or actual conflicts of interest during tenure, promotion, or other department proceedings gave rise to this restriction. We continue to think that it is important for faculty to have strong supportive professional relationships with senior people outside their department. Faculty without tenure must be more circumspect in their dealings with senior members of their department than with someone from the outside.

The sponsor may come from a department other than the associate's at Hunter, from any of the CUNY colleges or the CUNY Graduate Center, or from other universities in the New York area.

At many institutions, especially those whose primary commitment is to undergraduate teaching, a faculty member might be the only person or one of a very few people in his or her specialty. We have found that most of the research-intensive institutions from which faculty get their degrees or postdoctoral experience do not provide sufficient explicit skills and information about how to succeed in academic institutions or in their disciplines. People leave graduate school not understanding how academia works in general, how gender and race complicate achievement, or how to balance competing calls on their time. Most are also still novices at writing grants and papers, presenting colloquia and talks, and dealing with journal editors and reviewers. Many of their supervisors provided them with a minimum of attention and training.

At large research-intensive institutions, most faculty are in a community of highly productive peers, experience great pressure to become similarly productive, and have colleagues on hand with whom to collaborate or discuss their research. Location—one's place of work—determines productivity as much as or more than one's productivity determines one's location (Allison and Long 1990; Long and McGinnis 1981). Well-financed institutions tangibly and intangibly support the research enterprise.

The majority of institutions, however, are not well financed and provide only limited access to seasoned, senior investigators. Once a faculty member is at an institution with little research support, she or he will find it increasingly difficult to develop or maintain scientific relationships with others in the same specialty. Access to a sponsor can make all the difference between being on the outside and being on the inside.

Once accepted into the program, the associate works with her department chair, other senior colleagues, and the GEP codirectors to identify a possible sponsor. In some cases the associate already knows the possible sponsor and speaks informally to him or her. A GEP codirector officially approaches the sponsor on the associate's behalf, typically by a letter that describes the program, the sponsor's responsibilities, and the compensation. The sponsor must commit to a set of goals and activities, including biweekly communications with the associate, and—a recently added feature—one interview per semester with the GEP to discuss the associate's progress. The sponsor must be physically located within travel distance of Hunter College, in order to make it more likely that the sponsor and associate will meet face to face. (Originally, we hoped that sponsors would attend some of the workshops, and a few did so. The associates, however, were able to speak more freely when their sponsors were not present.)

The $5,000 sponsor stipend was estimated as the minimum attractive dollar amount for an accomplished, busy, senior person. We ask for considerably more intellectual attention than most mentors provide. Compensating people acknowledges their expertise and the value of their time. People take their work more seriously and perform it more reliably when they are paid. When one potential sponsor, who had already had some contact with an associate to advise her about a book manuscript, was approached the sponsor said, "Oh, now I'll really need to be serious about how she revises the manuscript so that the book will be published." That was exactly the response we wanted. The manuscript was revised many times and the book has been published by a strong university press. Other sponsors have also commented on the commitment implied by the financial benefit.

Another reason for our paying the sponsor was that we wanted the associates to feel entitled to call on their sponsors for substantive intellectual commentary and professional advice. Inculcating entitlement in some of our associates has been difficult. Some associates find it difficult to approach their sponsors despite the compensation. For example, one associate in her second year said that she *now* felt more comfortable about approaching her sponsor. She understood that he was being paid, and that helped, but it was still difficult for her to initiate an interaction. Associates' reluctance to take up their sponsors' time continues to be a problem.

Despite the compensation, not all matches work. About one-third of our sponsors worked out extremely well, about one-third worked out well, and about one-third did not work out well. We now tell sponsors at the outset that some associate-sponsor pairings work better than others and that at the end of each semester, in our discussion with them, we will consider whether the match seems to be a good one and should continue. We accordingly now pay sponsors in two installments.

Although we added the discussion with a GEP codirector with the intention of providing oversight and accountability, we have learned that sponsors enjoy and look forward to the discussions. They get new ideas, pass on their insights and progress, reflect on the structure of their profession, and feel part of an important project. Several of our sponsors from research-intensive institutions have commented on the value of the program and noted that their institution does not, but should, provide a comparable program. Pleasurable engagement leads to commitment.

Our construal of the role and importance of the sponsor has changed over time. We have begun developing the implications of our concept of a circle of advisors, also known in the literature under a variety of names—mosaic mentor, composite mentor, and the like. The idea of a circle of advisors is that any individual has a wide range of needs: information, advice, support, challenge, encouragement, critiques, advocacy, and so on. In the ideal case, each individual creates a list of what she needs and for each need tries to identify someone who could fill that need. Today's circle of advisors is not necessarily tomorrow's. As one's needs change, so will one's circle.

Incorporating this into the Sponsorship Program enlarges the kinds of models we will entertain. For example, we could *(a)* provide each associate with two sponsors, one for intellectual support and one for psychosocial support, or one for one aspect of the associate's research and one for another aspect, *(b)* facilitate collaborations between associates and nonlocal colleagues, *(c)* provide sponsors for some associates but not others, *(d)* encourage associates to form a circle of advisors without providing a primary sponsor, or *(e)* tailor the model to the individual needs of each associate.

The Workshops

New associates receive two days of intensive workshops during the summer after they are accepted into the program. Monthly workshops on issues of continuing interest are held during the fall and spring terms. Workshop topics cover the techniques, skills, strategies, and knowledge necessary for professional success, such as time management, grant writing, paper presentations, negotiation, developing useful contacts with colleagues, developing collaborations, identifying areas where technical skills are necessary, dealing with rejection, and understanding the role of gender as a determinant of professional success.

The workshops are arguably the most visible, transportable, and successful component of the Sponsorship Program. Many faculty, especially women and minorities, are intellectually and socially isolated. The MIT report (Massachusetts Institute of Technology 1999) specifically noted the marginalization that senior women in the school of science perceived. The workshops feed a hunger for information, guidance, and community, and as such have become a major focus of our efforts. We have learned that this hunger is widespread, and exists among men as well as women, nonscientists as well as scientists. We have successfully developed the workshops as a stand-alone feature, compressed them into one-, two-, or three-day formats, and adapted them for different groups within and outside CUNY. Following a suggestion by one workshop attendee, we are in the process of developing a book based on the workshops.

The list of workshop topics has grown over time, but most can be subsumed in a few broad categories:

career development: balancing work responsibilities, making effective public presentations, preparing one's vita, developing self-presentation skills, building a national reputation, creating a circle of advisors, handling power and politics, teaching effectively and efficiently, increasing negotiation skills, preparing for tenure and promotion, capitalizing gains and maximizing progress in the summer

writing and publishing: managing time and overcoming procrastination, publishing and handling rejection, writing grants

mentoring and leadership: being sponsored and sponsoring others, managing laboratories, research assistants, and students

balancing work and personal life: developing equality in personal relationships, balancing work and personal life

Throughout the workshops, we discuss how gender schemas and the accumulation of disadvantage affect women in the workplace (Valian 1998), and use social science theory and research to support claims and recommendations. Increasingly inspired by the comments of our associates, we also consider how culture and ethnicity play a role in professional success. Workshops are evaluated by all participants, and the evaluations have shaped their evolution.

No aspect of the Sponsorship Program has evolved more fully in form or content than the workshops. They began as twice-monthly meetings among the associates and GEP codirectors and staff. Each workshop was two hours long, and consisted of hour-long presentations by the codirectors (or occasional other experts) followed by open discussions among the entire group. To convey the evidence-based nature of the program, we assigned homework in the form of scholarly readings from the social sciences.

It quickly became apparent that associates wanted fewer scholarly readings, shorter presentations, and more opportunities to engage with their colleagues. We responded by assigning a smaller number and larger variety of readings, including practical guides. We developed "tips" sheets that summarized a great deal of information succinctly. We also assigned some preworkshop activities that set the stage for constructive, informed workshop activities and discussions. We shortened our presentations to no more than twenty-five minutes, and designed structured workshop activities for small groups of two to four so that associates could interact with and learn from each other. These changes have been enthusiastically received.

One workshop that has had a demonstrable impact focuses on tips and strategies for public speaking and presentations. One goal is to improve associates' presentations—to help them give talks that capture the audience's attention, clearly state the main points of the research, and explain why the research is important. Although these points are obvious, and already known by the associates, knowing and doing are two different things—as anyone who has sat through dozens of professional talks at conferences knows. We also teach associates how to receive feedback from their audience. A second goal is to improve how the associates listen to others' presentations and provide constructive commentary.

To prepare for the workshop, associates read a document we have prepared (developed with Nikisha Williams) on what should be presented in the first three minutes of a talk. The document emphasizes trying to find the right example or phenomenon to introduce a topic. Then the associates create their own three-minute opening. At the workshop each person gives her opening, which is timed. The listeners give no more than a minute's worth of comments.

Some associates "forget" that they were to prepare an opening and come to the workshop unprepared (but must give the opening, anyway). All associates learn something of value about how to improve their organization and delivery from each other's comments. The light goes on when the right opening fact or example grabs the audience and sets the tone for what follows. Associates also learn how to frame feedback that is specific and useful without being harmful.

Another particularly useful workshop asks associates to plan the summer so that they can take advantage of the absence of teaching and administrative work and make progress on their research. In one activity, associates work with a partner to design a summer research schedule that works best for them and come up with concrete suggestions for how to put such plans in motion. Would a visit to another lab be useful? What arrangements need to be made? Does a home office need to be equipped or rearranged? What needs to be purchased or borrowed? Are babysitters needed? How can they be lined up early?

A second activity for planning the summer occurs before the workshop. Associates talk with at least one colleague whose work habits and productivity they admire, and ask how they spend the summer: how do they structure their days and weeks; where do they work; what do they do about their research when they travel; how do they balance work with other responsibilities and leisure activities; with whom do they talk about their work during their summer? All of us learn that productive people have a range of strategies. For some, the summer marks a time of more intense research activity, with more interns and others working in the lab. For others, summer commitments differ little from the rest of the year: graduate students and postdocs require continuous supervision; laboratory research is ongoing. Some people work full-time (forty to sixty hours per week) on their research. Some nonlaboratory scientists come to their office every day because they work more effectively at their office than at home. Some have working vacations in which they work in the morning and have fun in the afternoon. Others deliberately do no work at all on vacation. Yet others use part of their vacations and family visits for scholarly reading. Learning what productive people expect of themselves helps associates think about their own productivity and the balance between work and leisure that they want to have.

A constant refrain from our associates is that they wished they had had various forms of information or guidance sooner. In response, we have compressed the workshops for all new associates, and offered all of the basic topics in condensed form in a two-day, intensive format at the beginning of the summer of their first sponsorship year. In this way, we are able to start people's summers off in a productive way. We continue to have more advanced monthly workshops during the fall and spring for all the associates.

Relatively inexpensive and high impact, the workshops are an excellent candidate for institutionalization and dissemination. In May 2004 we offered a three-afternoon set of workshops for women in science across CUNY: it was a resounding success. The only request for change was to add a dinner at the end so that people could talk with each informally. We received a CUNY Faculty Development Award to host a two-day, sixteen-hour workshop for male and female junior faculty in any discipline across CUNY in December 2004. Once again, the response—both faculty interest (we had a waiting list of twenty-one after our cap of thirty) and evaluations of the workshops—was overwhelmingly positive. We received funding from the Office of the Executive Vice Chancellor of Academic Affairs, CUNY, to hold the workshops again in May 2005. Finally, in collaboration with the New York Academy of Sciences in spring 2005, we held three workshops at Hunter for graduate students, postdocs, and junior faculty in natural science. The sixty people who attended came from a wide range of universities in New York, Connecticut, and New Jersey; they too provided glowing evaluations. Colleagues from other CUNY campuses are now training with us to learn to conduct similar faculty development workshops at their home campuses.

Consultations with the GEP Codirectors

A GEP codirector speaks "officially" once a semester to each associate and is also available whenever associates ask for advice on topics that range from preparing tenure and promotion packages to negotiating for better teaching schedules to responding to rejection. We had not anticipated how important the consultations would be. The GEP codirectors act as general advisors to associates on topics ranging from intellectual and scholarly matters to institution and department issues to interpersonal conflicts on the job and work-family conflicts at home. We discuss how they can succeed at Hunter College and in their disciplines. We are generally accessible to associates and regularly communicate with them in person, and via phone and email. In their ratings of the importance of various components of the Sponsorship Program, associates strongly disagreed with the statement, "The Sponsorship Program would be just as valuable without VVV and VCR" ($M = 1.73$, $SD = 1.14$, where 1 means strongly disagree and 6 means strongly agree). Our growing role in the advisement of associates has contributed to our understanding of the desirability of a circle of advisors, rather than a single sponsor, to be discussed further below.

Measurement

The major types of measures are evaluations of components of the Sponsorship Program, particularly workshops, associate progress reports, and associate interviews with the GEP codirectors. Regular measurement serves many purposes in the Sponsorship Program, including education, feedback, and evaluation. The associate progress reports have undergone the most refinement and best illustrate the education, feedback, and evaluation functions of measurement in the Sponsorship Program.

We make use of both qualitative and quantitative progress reports, and ask for some information monthly (scholarship and sponsor interactions), and some three times a year (teaching, service, and student or lab supervision). For monthly reports, we use a comprehensive, quantitative survey of scholarly activities (for example, work on refereed journal articles, internal and external grants, and other writing; professional development and other activities) and sponsor interactions, followed by narratives that cover such topics as what barriers associates have encountered and how things can be improved. The teaching and service survey asks associates to describe the courses taught, people supervised, and service to the profession, the college, and the community.

Over time, we have made the monthly checklist more comprehensive and easier to use. In order to reduce the reporting burden on associates who are no longer funded by the GEP, while still allowing us to track their progress in scholarship, teaching, and service, we recently developed a new checklist. Associates who are not currently funded are asked to complete this checklist three times a year (after each semester and at the end of the summer) for the two years following their GEP funding.

We created an impact survey to capture large program effects. This survey asks associates to tell the GEP about (a) the number of undergraduate, masters, doctoral, and postdoctoral students they have supervised, (b) the professional activities of those students (i.e., conference presentations, manuscript submissions, articles published), (c) collaborations that have developed between associates and sponsors or associates and other colleagues, (d) whether Sponsorship Program participation has affected the associates' relationships with students and their likelihood of serving as an advisor to a student or colleague, (e) whether their participation in the program has resulted in more active participation in departmental, college, university, or discipline-wide activities to increase equity, and (f) their leadership activities. We also created a survey to evaluate the relative importance of various program components (e.g., funding, the sponsor, advice from the codirectors, workshops, interactions with other associates).

Associates are not uniformly vigilant in submitting their monthly reports, especially when they have little progress to report, and we have worked to streamline reporting to remove impediments. But, overall, associates find the process of filling out progress reports instructive and motivating. The reports remind them of their commitments, help them to break down large projects into smaller tasks, identify barriers to achievement, and keep the focus on academic writing.

Evaluation of the Sponsorship Program

Because the chief goal of the Sponsorship Program is to increase scholarly productivity, the crucial test of its effectiveness is whether participation increases paper and grant submissions. Dependent t-tests determined whether associates' paper and grant submissions increased as a result of their participation in the Sponsorship Program. The sixteen associates in cohorts 1 and 2 submitted significantly more papers and grants (internal and external) during their first year in the program ($M = 5.5$, $SD = 3.5$) than they did during the year before entering the program ($M = 3.1$, $SD = 1.8$), $t(15) = 2.32$, $p < .05$. The eleven associates in cohort 1 presenting full data also submitted more papers and grants during their second year of the program than they had during the year before entering the program, $t(10) = 3.2$, $p < .01$. Thus, in a relatively short period of time, program associates not only increased their productivity but also maintained it. Similar significant differences exist when only paper submissions are examined.

The positive effects of associates' increased productivity affects others at Hunter, including students. Of the total group of twenty-four associates who have been in the program, eleven have provided information about work with students: they have supervised a total of forty-five undergraduates, thirty-four master's students, and twelve Ph.D. students. The students have presented their research at conferences (22 posters, 23 papers, 3 invited talks, 8 symposia) and have been sole or coauthor on three papers in refereed journals. Eleven of the undergraduates have gone on to M.A. programs; ten of the undergraduates and M.A. students were accepted into Ph.D. programs.

Although we lack desirable data from comparison groups that would allow us confidently to attribute the associates' performance to the Sponsorship Program, the qualitative assessments suggest a causal relation between being in the program and increasing productivity and commitment to research.

Qualitative Assessments

Some types of data can only be acquired through interviews and semi-structured discussions. One of our concerns is how to measure intellectual development, the growth of intellectual aspirations, and increases in intellectual engagement. We want to ensure that women scientists flourish intellectually, that they are able to develop their ideas and get satisfaction from their intellectual work. But what do we know about the intellectual life-cycle? What do we know about what contributes to intellectual growth?

We see women's high intellectual aspirations, their intellectual success, and their influence over the direction their field is taking, as signs that initiatives like the Sponsorship Program are on the right track—but all of these are very hard to measure. We want to understand the nature of intellectual development and intellectual influence. We want to understand what helps someone develop her (or his) intellectual and scientific contributions and what helps someone have an impact on her (or his) field. It is difficult to quantify such changes in an individual and even more difficult to understand what the causal factors are in helping someone bring out her or his best.

Our interviews with associates usually stimulate them to think about their progress—or lack thereof. We include excerpts (edited for clarity, ellipses not included) from letters they have written us.

The first is from an associate who was outside of academia for several years after getting her degree:

> Although the "measurable effects" of my participation in the GEP may not seem spectacular measured in terms of publications, I would like you to know that the GEP has played an enormous role in my determination to work to keep this job. I enjoyed working at Hunter my first year, but the next two years were often unpleasant. I questioned whether this was the right "work" given my unhappiness. I was not sure how much longer I wanted to stay.
>
> I cannot fully explain the role [the] GEP played in changing my mind about work, but it was significant. Three elements come immediately to mind. First, simply having to go to the city to attend meetings and report in kept me thinking about my work, when I otherwise might have abandoned it. Second, the GEP sessions encouraged me to focus on what it was about this career that

I did like, and what I would greatly miss without this specific job. Virginia's article [Valian 1977] on the nature of "work" made a very strong impression. And third, participation in the GEP is how I discovered both that there is a Hunter College community outside my department, and that the outside intellectual community, from which I felt so estranged, would accept me again if I made the effort.

The bottom line is that while I may not do enough, soon enough, to get tenure, I have already enjoyed work more this year than any other.

Another associate who, after a period of inactivity, applied for six grants during her two years in the GEP wrote:

As I look back at the progress I have made since the fall of 2003, I can see that without the GEP I would still be feeling depressed, overworked, and unaccomplished. All these feelings were swept away during the first year of my participation in the GEP seminars. Understanding that my feelings were shared by others, and that they were essentially a result of my having been disconnected from my work, was an immense relief. I would like to thank the GEP for what it has done for me, and helped me to do for myself.

A third associate wrote:

I successfully prepared my promotion package, had my interview, and have been promoted. My promotion process apparently went very well. Furthermore, I now have also finished my part of the tenure process, and gone again successfully through the interview. Although it is early to tell, I know that the endorsement for me was unanimous. I will never be able to thank the GEP program enough for how I approached promotion and tenure, and how well I did through all of it, as well as for the financial support. Most important for me: being part of a program that made me stay "on track" and focused.

A week before I had my promotion interview I had a very nasty rejection of a paper of mine from a journal—that was an incredible setback at the time. I think that my experience with the GEP helped me deal with it in a professional manner and certainly made it easier for me not to let this spill into preparation for the interview. I put it aside and concentrated on what I had, *not* what I did not! Then I began the process of thinking about another journal, settled on a top journal, revised the manuscript to meet their main purpose and specifications, and submitted a much better article to this journal, which was accepted with minor revisions.

I have taken more initiatives at the department level and as a result people listen to me more. I was asked to serve as the chair of an important departmental committee. In this committee, one of the persons who had provided the most obstacles for me throughout my years at Hunter now defers to my authority as a chair. I consider this an incredible accomplishment.

We know from our quantitative data and from our interviews that a minority of associates continue to be less productive than they or we hoped would be the case. Nevertheless almost all associates see themselves as further ahead than they were when they started, even if there is little tangible evidence to back that up.

We also see that our associates help other faculty. They pass on to their junior colleagues what they have learned about how the institution works, how their discipline works, how to be more effective within their department, how to improve departmental governance. Because they learn about how other departments are run by trading experiences with other associates, they see that they can have a hand in making their department a better place for faculty.

When queried about the effectiveness of our program, some associates indicate that they principally needed time and money, although they benefited—albeit less directly—from the workshops. Others talk about how important their sponsors have been: direct intellectual feedback, collaborations that have developed as a result of the sponsorship, meetings with other people that have developed. Yet others specifically say that financial support was less important than the other forms of support—support from the codirectors and the GEP as a whole, support from workshop content and discussion, support via an increased sense of belonging, support simply from having been chosen as someone the GEP wanted to invest in. Some mention every aspect of the program and think that the package is important. What we have learned from our interviews, discussions, and correspondence with our associates is how rich—and untapped—an area intellectual development is.

The Sponsorship Program is how the GEP is best known at Hunter, and the College recognizes and acknowledges its benefits. At College-wide tenure and promotion proceedings, membership in the Sponsorship Program is frequently credited with contributing to a candidate's success, and is increasingly touted as one of a candidate's' achievements in and of itself. This suggests that the program is perceived by department chairs and top administrators alike as effective and prestigious. Through the Sponsorship Program's success, Hunter's president, Jennifer Raab, became convinced of the importance of increased opportunities for faculty development for scholarly productivity, and is working with the GEP codirectors to institutionalize the program and develop future funding. She has also provided "step-down funding" for a subset of third-year associates who have made demonstrable progress but require additional funding to complete projects.

Leading Ideas from the Sponsorship Program

The following six principles or leading ideas continue to guide the refinement and expansion of the Sponsorship Program and other initiatives. Grounded as they are in the literature on interventions in organizations and experimental work in the social sciences, especially social psychology, these principles can be applicable to other programs and institutions.

1. *Gender is a window on institutional effectiveness.* When attention is paid to gender issues and sex comparisons and disparities, problems in an institution that are totally unrelated to gender become apparent. When we attended to problems of faculty women scientists at Hunter, it became clear that there was a general need for faculty development for research and scholarship.

2. *A continuous thread links undergraduates, graduate students, postdocs, and faculty.* The connection between faculty and students is critical for improving the representation, retention, productivity, and morale of women in the sciences. Students at all levels benefit from seeing and interacting with female faculty whose science careers are thriving and who maintain a desired balance between careers and personal lives. Female students are more likely to enter and continue in the field if they see that scientists can "look like them." The increased research activity of our associates has spilled over to our undergraduate, master's, and CUNY doctoral students, who now have more opportunities for collaboration and support.

3. *Women who profit from initiatives like the Sponsorship Program are carriers of information and strategies to colleagues in their departments and throughout the college.* Women who benefit from program initiatives can become seeds of change in their departments and other groups by sharing materials, increasing their influence, nominating female colleagues for honors, awards, and opportunities, serving as advisors to others, and working with administrators to improve conditions for women. Several associates have started building alliances within their departments and programs, vouching for each other, pooling resources, and collaborating on research and other projects at Hunter

4. *A circle of advisors is superior to a single mentor.* No one mentor can or should provide the myriad intellectual, professional, and psychosocial supports that junior faculty need, and from which all faculty can profit throughout their careers. A circle of advisors minimizes the nature and extent of any one advisor's responsibility, makes it more feasible to attract advisors in the first place, and maximizes the chances that faculty will get the help they need from the best possible sources.

5. *Measurements, interviews, and application procedures are interventions.* Targeted interviews can be a more efficient way to collect information than developing comprehensive surveys. Targeted interviews with representative groups can elicit the same information as comprehensive written surveys, and, because they are open-ended, elicit new information not anticipated by researchers. Interviews can lead directly to solutions, as the process of discussing issues causes people to engage more fully and consider topics more thoughtfully (Valian and Fletcher 2004).

6. *Attention to gender encourages distributed leadership.* Long-standing institutional change with regard to gender equity and diversity requires shared goals and the development of consensus on policies and procedures. Enduring change is most likely to happen when leadership is distributed throughout an organization, and people at all levels are, and perceive themselves to be, responsible and effective agents of change.

REFERENCES

Allison, P. D., and J. S. Long. 1990. Departmental effects on scientific productivity. *American Sociological Review* 55:469–78.
American Association of University Professors. 2005. Inequities persist for women and non-tenure-track faculty: The annual report on the economic status of the professions, 2004–05. *Academe* 91.2: 20–30.

Cataldi, E. G., E. M. Bradburn, and M. Fahimi. 2005. 2004 National study of post-secondary faculty (NSPOF:04): Background characteristics, work activities, and compensation of instructional faculty and staff: Fall 2003. Washington, DC: National Center for Education Statistics, US Department of Education. http://nces.ed.gov/pubs2006/2006176.pdf, accessed December 22, 2005.

Levy, R. L., D. Yamashita, and G. Pow. 1979. The relationship of an overt commitment to the frequency and speed of compliance with symptom reporting. *Medical Care* 17 (3): 281–84.

Long, J. S., ed. 2001. *From scarcity to visibility: Gender differences in the careers of doctoral scientists and engineers.* Washington, DC: National Academy Press.

Long, J. S., and R. McGinnis. 1981. Organizational contexts and scientific productivity. *American Sociological Review* 46:422–42.

Massachusetts Institute of Technology. 1999. A study on the status of women faculty in science at MIT. *MIT Faculty Newsletter* 9 (4). http://web/mit/edu/fn1/women//women.html.

National Science Foundation. 2004. Women, minorities, and persons with disabilities in science and engineering, 04–317. http://www.nsf.gov/sbe/srs/wmpd/pdf/front.pdf., accessed March 30, 2005.

Valian, V. 1977. Learning to work. In *Working it out: 23 women writers, artists, scientists, and scholars talk about their lives and work,* ed. S. Ruddick and P. Daniels, 162–78. New York: Pantheon.

———. 1998. *Why so slow? The advancement of women.* Cambridge: MIT Press.

Valian, V., and J. Fletcher. 2004. Personal communication.

Wurtele, S. K., A. N. Galanos, and M. C. Roberts. 1980. Increasing return compliance in tuberculosis detection drive. *Journal of Behavioral Medicine* 3:311–18.

EIGHT

Supporting Faculty during Life Transitions

Eve A. Riskin, Sheila Edwards Lange, Kate Quinn,
Joyce W. Yen, & Suzanne G. Brainard

ONE OF THE most-cited advantages of a faculty career is its flexible schedule. As long as they are productive, faculty members can set their own work schedules and, in some cases, work locations. However, it has been demonstrated that this flexibility blurs the boundaries between work and life and, typically, results in *less* time for the personal lives of faculty (Bailyn 1993; Hensel 1991; Sorcinelli and Near 1989; Williams 2000a, 2000b; American Association of University Professors 2001). Consequently, the same flexibility that was an advantage can become a burden as a personal crisis can easily consume a tremendous amount of time.

Studies on faculty stress have identified lack of time, family needs, and pressure to do the research needed for promotion as the most common sources of stress among university faculty (Barton, Friedman, and Locke 1995; Gmelch, Willse, and Lourich 1986; Jacobs and Winslow 2004; Grant, Kennelly, and Ward 2000; Thompson and Dey 1998). For many faculty members, a substantial amount of the time spent on campus is devoted to teaching and service, while evenings and weekends are used for research and writing. Children, major illness, hidden disabilities (Bertz 2003) or caring for an elderly parent all detract from the time available for research and writing. Often the time used to attend to major transitions in personal lives would otherwise be spent on research needed for promotion and advancement.

"Spillover" between the time demands of a faculty career and family responsibilities has a high correlation with faculty satisfaction with both life and work (Sorcinelli and Near 1989) and causes more stress for women faculty than it does for men faculty (Davis and Astin 1990; Wilson 2001; Thompson and Dey 1998; Gatta and Roos 2002). Although a supportive spouse or partner can alleviate some of the stress, women still bear most of the responsibility for caregiving, and men and women make very different trade-offs in how they allocate their time (Valian 1998). Mason and Goulden (2002) found that faculty career transitions that coincide with child-bearing are particularly costly to women's advancement and promotion. According to their study, women who had babies within five years of earning their doctorates were less likely to earn tenure than men who fathered babies in the same time period. The gap in tenure attainment between men and women with early babies is 24% in the sciences and 20% in the humanities and social sciences.

Given that the average age of granting a Ph.D. is thirty-three (Mason and Goulden 2002) and the average age of all assistant professors is forty years old (Jacobs 2004), an increasing number of faculty can no longer afford to delay family formation until the attainment of tenure (Armenti 2004; Blackburn and Hollenshead 1999; Gatta and Roos 2002; Varner 2000). Recognizing the impact of midcareer life transitions on faculty, the American Council on Education (2005) published *An Agenda for Excellence: Creating Flexibility in Tenure-Track Faculty Careers,* which calls on universities to find creative ways to permit flexibility in faculty careers that will facilitate life transitions.

The nature of faculty work makes transitions difficult regardless of gender. Nevertheless, women faculty members tend to shoulder much of the work surrounding transitions. Moreover, faculty in science and engineering fields face an additional challenge due to the nature of their research. Science and engineering research is often time-sensitive and may require significant start-up effort to establish results. Major disruptions in productivity caused by difficult transitions can result in a loss of research results or momentum that can be career-ending in science and engineering. The gender disparity in experiencing transitions, coupled with these characteristics of science and engineering research, provide the context for programs to support faculty undergoing transitions. The initiatives described here grew out of programs to support women faculty in science, engineering, and mathematics. Recognizing that transitions affect all faculty, though, programs to support faculty encountering difficult transitions should benefit all faculty, regardless of gender. These programs provide another example of how initiatives initially conceived to help those who are differentially disadvantaged result in improved resources for all.

As part of its National Science Foundation (NSF) ADVANCE program, in 2001 the University of Washington (UW) established a Transitional Support Program (TSP) to alleviate the effect that major life transitions have on faculty careers. The TSP awards grants of $5,000 to $38,000 to outstanding science, engineering, and mathematics (SEM) faculty in the midst of major life transitions, such as the birth or adoption of a child, personal medical needs, family illness, or caring for an elderly parent. The TSP also helps faculty members who need a modest level of support to advance from assistant to associate professor, associate to full professor, or into positions of leadership. The primary focus of the program is to help faculty as they deal with the stress and challenges of balancing an academic career with a personal life.

What follows is a synopsis of UW's ADVANCE program; a description of how the TSP was implemented; a summary of evaluation results; an overview of similar programs at other institutions; and the outlook for institutionalizing the program after the grant period ends. Recommendations on how best to replicate the program at other institutions are also provided.

UW ADVANCE

The University of Washington was one of the initial nine universities selected for a National Science Foundation ADVANCE grant in the fall of 2001. The UW has a Carnegie ranking of Doctoral/Research University/Extensive and for the past thirty years it has been the number one public university in the country in receiving federal support for research and training. Its female science and engineering faculty members are recipients of numerous national awards, including NSF Young Investigator awards, NSF CAREER Awards, Presidential Early Career Awards for Scientists and Engineers, and Sloan Research Fellowships. The UW ADVANCE project was designed to build upon existing strengths at the university while serving as a catalyst for institutional transformation. The project's vision as articulated in the original proposal is "a campus in which all SEM departments are thriving, all faculty are properly mentored, and every SEM faculty member is achieving to his or her maximum potential."

The UW ADVANCE program created a Center for Institutional Change (CIC) that is advised by a joint leadership team from the College of Engineering and the College of Arts and Sciences. The leadership team includes deans, chairs, and faculty members from both colleges, the executive vice provost, the vice president for minority affairs, and the vice president for human resources. Academic departments participating in the program include all ten within the College of Engineering and all eleven departments in the College of Arts and Sciences' Science Division.

The number and percentage of female ladder faculty as of 2004 (in the original nineteen ADVANCE departments) in each college are shown below in table 1. Ladder faculty at the University of Washington include tenure-track faculty and a special category of faculty who are without tenure by reason of funding (WOT).

Implementing the TSP

The Transitional Support Program (TSP) at the University of Washington was implemented in fall 2001. From the beginning, it was decided that the program would be competitive; a variety of circumstances and proposed remedies would be considered; the selection process would be managed solely by the CIC director and the ADVANCE principal investigator; the request for proposals would be offered multiple times per year; and participation in an evaluation process would be mandatory for all recipients. The program is open to both men and women faculty in ADVANCE departments who are undergoing transitions.

In an effort to avoid perceptions that the program is remedial, the ADVANCE leadership team decided upon a competitive application process. It was hoped that the competition would signal to recipients and their colleagues that the award is given only to outstanding faculty members who are deserving of the honor. To enhance the prestige of the award, recipients are called ADVANCE Professors. Applicants must submit written statements about the type of transition they are experiencing, how it has impacted their career thus far, and how an award will

TABLE 1. UW ADVANCE Ladder Faculty, Fall 2004

	Total	Female	% Female
College of Engineering	206	33	16.0%
College of Arts and Sciences (Science Division)	255	35	13.7%
Total UW ADVANCE	461	68	14.8%

help to advance their career during the transition. A letter of support from the department chair is strongly encouraged. If, however, the applicant indicates that lack of support from his or her chair has affected the applicant's ability to endure the transition, a letter of support is not required.

Through spring 2005, thirty-five faculty members from participating SEM departments have been designated as ADVANCE Professors. TSP grants totaling $727,443 (including UW matching funds) have been awarded to faculty members in participating SEM departments. While most awards have been given to female faculty members, two awards from UW matching funds have been given to male faculty members with a significant family responsibility or crisis. Recipients reflect the diversity of faculty in participating departments, and include four Asian/Pacific Islander, one African American, and one Latina/o faculty members.

As shown in figure 1, the transitions experienced by ADVANCE Professors have been as varied as the remedies proposed.

Faculty are advised not to include information about a medical condition in the TSP application but to work with the UW's Disability Services Office to obtain a letter affirming that a qualifying medical condition exists. Faculty medical leave must be used first in the case of personal medical conditions. If a faculty member's transition situation changes unexpectedly (i.e., a failed adoption or pregnancy), the award is reevaluated and the award is adjusted as appropriate to the new situation. All applications are confidential, and the CIC director is extremely discreet about who is consulted when seeking remedies.

As applications are reviewed, the CIC director works very closely with the applicant and his or her department chair to develop cost-effective remedies. Graduate or postdoctoral student support has been one of the most popular remedies for faculty who need more time for research. One significant outcome of the CIC director and department chair collaboration on remedies has been the leveraging of award funding. Many of the department chairs have matched TSP awards with in-kind support, cash contributions, or course releases to provide the ADVANCE Professors additional time for research.

Decisions on TSP awards and remedies are left solely to the CIC director and the UW ADVANCE principal investigator. The autonomy and flexibility in decision making enables the director to negotiate remedies with department chairs with little bureaucracy. Once a deci-

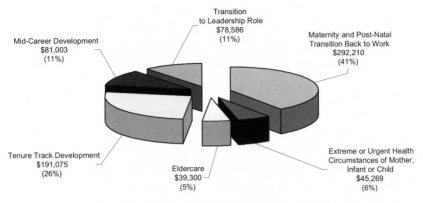

Figure 1: TSP Funding by Categories, 2001–5

sion to grant an award is made, very little administrative action is needed to execute the negotiated remedy.

Another benefit of the tight decision-making process is that the request for proposals can be offered three times per year with deadlines in November (support for winter quarter), February (support for spring quarter), and May (support for summer and fall quarters). Male and female faculty members in UW ADVANCE are sent quarterly email announcements about the program and its deadlines. The UW ADVANCE web site has deadlines listed for each quarter to allow faculty to plan ahead and a budget tool to help them calculate the costs of release time or graduate support.

Applicants are notified of their selection by the CIC director and reminded that they will be required to participate in the program's evaluation process. The evaluation is managed by the Center for Workforce Development at the University of Washington. The evaluation is structured to provide feedback to program managers that is used to improve program management and delivery. The evaluation consists of an initial interview with all recipients and annual follow-up emails requesting a statement on the impact that the award has had on their research productivity and career advancement.

The TSP has been able to fund most requests that fit within the broad definition of life transition, which includes health or family-related transitions, transitions to new career positions, and transitions to positions to leadership. Requests for sabbatical support or to write textbooks were not funded. In cases where the application did not fit the program, the CIC director phones the applicant directly. In several cases, the proposal was reframed and then funded in a later round.

While it could be useful to have more input on funding decisions, using a smaller review committee ensures that memory remains in the system and that the grant recipients receive their funding very quickly.

The TSP program at the University of Washington has a narrower definition than some other grant programs offered at other institutions. (See the "TSPs at Other Institutions" section for descriptions of other programs.) While more broadly defined grant programs can address a variety of issues, the UW program is targeted more narrowly to highlight the importance of supporting faculty undergoing transitions. Doing so recognizes that faculty lead multidimensional lives, and that these dimensions are not compartmentalized. Faculty productivity is affected by such personal transitions. Moreover, at the University of Washington, support of this type was typically elusive unless a department chair or dean knew how to navigate the institution to obtain financial support for critical faculty transitions. A narrowly defined program such as this one provides a clear recognition of the institution's commitment to supporting faculty, and releases the burden of justifying how critical events in a faculty member's life affect his or her work.

Evaluation Results

As part of the program evaluation, all ADVANCE Professors are interviewed one year after the receipt of the grant, and asked to complete annual follow-up surveys. The surveys collect information about milestones achieved, publications, research grants or leadership positions obtained since receipt of the grant. In addition to collecting the same kind of information about productivity, the interviews provide an opportunity for recipients to reflect on how their colleagues responded to the award, any negative repercussions related to asking for help during the transition, and whether the academic department changed how it handles transitions as a result of a professor's experience. Although there are currently thirty-five recipients, three have left UW and eleven report that they are not at a point at which they can evaluate the impact of the award.

Interviews and surveys from the remaining twenty-one recipients indicate that the program has greatly contributed to their productivity.

A brief summary of achievements is as follows:

- Six ADVANCE Professors who were assistant professors at the time of their award have been promoted to associate professors. Three who were associate professors have been promoted to full professor.
- Eight ADVANCE Professors have taken on new positions of leadership, including director of research center, leadership of editorial board, co-chair of a national conference, service on national committees, and chair of a discipline society section. This result is in addition to the four TSP grants given specifically to support transitions to leadership positions.
- A total of sixty-five publications were produced by the twenty-one respondents after receipt of their TSP awards.
- Fourteen of the twenty-one (67%) have been awarded new research grants since being named an ADVANCE Professor.
- Three ADVANCE Professors have received NSF CAREER awards, two of whom were awarded TSP grants to support their transitions back to work after maternity leave.

Interviews with the initial set of recipients provided important feedback about how department climate and the chair influence implementation of the grant, and how the faculty members themselves attempted to structure the transitional support in a manner to avoid bias from colleagues. Most recipients indicated that the rationale for the funding they received was invisible to their colleagues because of the discretion and understanding of the department chair. Recipients who described their chair and department as very supportive reported no concerns about negative repercussions.

Negative repercussions were reported by two recipients as an outcome of poor communication on the part of the chair, or other faculty members believing that the recipient was not carrying a fair share of the departmental load. While others did not report negative repercussion, two additional recipients reported taking action to avoid bias. These recipients intentionally focused on the need to structure their request for assistance in such a manner that departmental duties were not shifted to their faculty colleagues. They asked for research personnel rather than course releases as a means to avoid negative repercussions from colleagues. One additional recipient indicated that the transitional support was not viewed as negatively as stopping her tenure clock.

An unintentional consequence of the TSP has been the informal mentoring of those who submit funding requests, regardless of whether or not the request is selected for an award. When applications have not been selected for a TSP award, the CIC director has introduced the applicant to other campus resources and programs. In a few cases, the CIC director has been able to work with the applicant and other managers to solve the transition concern with little or no funding.

Postgrant Implementation

The NSF ADVANCE project goal is to transform the institution rather than provide programs that terminate with the grant funding. Thus the project has undertaken several strategies to create lasting impact.

In June 2004, the CIC, in cooperation with the University of Washington's Office of the Provost, extended the TSP to the entire UW campus, an important step in institutionalizing ADVANCE's programs and activities. The UW-wide program is much more restrictive than the ADVANCE TSP. It offers financial support to help outstanding faculty maintain their productivity while experiencing potentially career-threatening crises such as severe personal illness extending beyond the standard leave; severe or acute family illness; childbirth complications; or other critical situations. The funding is for course release or research support or both. The goal is to help these faculty members avoid losing ground in their research during times of crisis. Unlike the ADVANCE TSP, the new program is also available to research faculty and WOT (without tenure) faculty. Investing a small amount of resources in this manner makes economic sense; the cost of having a productive faculty member stop conducting research can be huge. In addition, the program sends a message to the faculty that the University makes a priority of helping its faculty balance work and life.

As previously noted, for faculty in science and engineering, an extended period away from research because of conflicts between work and family can end a career if grants run out, students find other advisors, lab space is taken away, and so on. To help sustain and institutionalize the ADVANCE TSP for science and engineering faculty, the ADVANCE team has initiated conversations with College and University Development officers to investigate support from foundations.

In addition to the institutionalization of the program, the CIC director provides chairs and deans with feedback about the impact of the program on faculty productivity. In quarterly leadership development sessions, decision makers are reminded of the leverage provided by the relatively inexpensive TSP awards. Chairs have been encouraged to develop similar remedies for potentially terminal associate faculty members.

Evaluation results are being used to inform institutional policy changes related to transitions. In the second year of the ADVANCE grant the UW received an Alfred P. Sloan Foundation Grant (Grant #2003–5–3 DLC) to examine part-time tenure-track models for faculty members. The grant allowed the institution to review existing procedures and craft a flexible policy for faculty during transitions. The ADVANCE evaluation process has also identified concerns about the policy at UW regarding stopping the tenure clock. Faculty expressed reluctance to delay the tenure process because they were concerned that their relationship with departmental colleagues would be harmed. This issue was discussed at length during a quarterly leadership workshop with department chairs, deans, and emerging leaders; and chairs now recognize the need to be proactive about creating flexible solutions to transitions. For example, chairs are encouraged to ask each junior faculty member, regardless of gender, at his or her annual review meeting if something happened that year that would make him or her eligible for a tenure clock extension.

TSPs at Other Institutions

Among the nineteen ADVANCE institutions in the first two rounds, at least five have implemented transitional support programs similar to the one at the University of Washington. Examples can be found at the University of Michigan, Utah State University, the University of Wisconsin–Madison, the Earth Institute at Columbia University, and Case Western Reserve University.

The University of Michigan's ADVANCE program created the Elizabeth C. Crosby Research Fund to help meet career-relevant needs of individual instructional track faculty in science and engineering if meeting those needs will help increase the retention or promotion of women scientists and engineers. The funding is available to support a range of scholarly activities, including expenses for specialized child-care when it affects the ability to continue scholarly activities.

Utah State is piloting a transitional support program for female faculty members in the colleges of Agriculture, Engineering, Natural Resources, and Science. The program is open to tenured and tenure-track women whose research has been delayed or interrupted (or who anticipate an interruption) by family or other responsibilities, such as childbirth, adoption, elder care, or divorce. The overall goal of the program is to help female faculty keep their research programs on track through difficult periods and to reestablish research programs after interruptions. Finally, the department must provide matching support directly through matching funds or indirectly through reduced work assignments.

The University of Wisconsin–Madison implemented a Life Cycle Research Grant Program, now called the Vilas Life Cycle Professorship program. The funds are available to faculty and permanent PIs who are at critical junctures in their professional careers when research productivity is directly affected by life events. Faculty may apply for varying amounts and academic purposes. The support may be used for release time, research support, conference travel, or other needs and lasts up to one year. The Life Cycle Research Grant Program was institutionalized in May 2005 through a $6.4 million endowment from the Vilas Trustees.

The Earth Institute at Columbia University has implemented transition support grants to assist women scientists in maintaining their research productivity during common transitions. The grants are not intended to provide supplemental support for family care. They provide partial support for a postdoctoral scientist, research assistant, or adjunct professor to assist women scientists in their ongoing research projects or take over teaching responsibilities so that a faculty member can focus on her research. The awards are intended to ensure that the research does not diminish and that women scientists can continue their research when limited by family or life transitions.

Case Western Reserve University has an ADVANCE Opportunity Fund to maximize the chance of success for women faculty by providing support for projects and activities where funding is difficult to obtain through other sources. All women faculty, including instructors and research faculty, are eligible to apply. Opportunities include seed funding for research, bridge funding when ongoing research funding has been suspended, grants to support writing of books, travel grants, child care costs to attend a professional meeting, or grants to conduct research at another institutions.

Recommendations for Replication

The TSP program at the University of Washington has been implemented to be flexible and responsive to the needs of faculty members. A variety of remedies have been used, and recipients have included male faculty members. Program evaluation indicates that in each case, the TSP recipient has increased his or her research productivity. Based on the experiences encountered during the implementation of the program at UW, the following recommendations are offered to institutions interested in replicating the program.

1. Be sensitive to faculty members who are concerned about bias from colleagues and those who want to structure awards to avoid such bias. When necessary, help faculty members craft remedies that do not create the impression that their workload is being transferred to other faculty in the department.

2. Consider adaptability of the model to the scale and culture of the institution. The program can be implemented at the department level or university-wide depending on interest and the amount of funding available. The culture of the institution might influence the scale of the program as well as the population eligible for participation.

3. Include chairs and deans in the planning and development stage. Chairs and deans are more likely to support a program when they are involved in its design. Their involvement will also lessen the need to explain goals and justify requests for their assistance in the development of remedies.

4. Build flexibility into the language used to solicit proposals. Rather than limiting transitions to specific criteria, flexibility will allow the program to respond to different transitions that faculty may experience.

5. Strongly encourage or mandate that applicants include a letter of support from their department chair in the application materials. When chairs provide a letter of support, they are forewarned about the type of transitional assistance needed. The letter is also a clear indication that the chair will be amenable to finding a solution to the faculty member's concerns.

6. Communicate clearly with all parties about how the award will be implemented. Unclear communication contributes to negative repercussions in faculty members' relationship with their chair and colleagues.

7. Include language that allows for a reassessment of the award if the cir-

cumstances of the transition change substantially from those described in the application. The ability to increase or decrease awards will allow the program to better respond to faculty needs. Changes in circumstances such as unexpected assistance from a family member, a miscarriage, a delay in adoption plans, or quick recovery from illness could change the type of transitional support needed during the award period.

8. Be cautious about restricting eligibility to women alone. Major life transitions affect both men and women, and in many dual-career families the man shares equally in caregiving responsibilities.

9. Establish an expectation that recipients will assist in the evaluation process. Doing so sends a message that a tangible outcome is expected at the end of the grant period. Further, faculty members are aware of the importance of accountability for program funds and know ahead of time that they must cooperate with program evaluators.

Summary

The presence of a transitional support program has been extremely valuable for faculty productivity and morale. During the evaluation process, many of the TSP recipients noted that for the first time in their careers they felt as if the institution were invested in their success. The availability of the program acknowledges that the institution values balance in faculty lives. The program at the University of Washington serves as a model for other college and universities seeking ways to support faculty during difficult transitions.

NOTES

This chapter draws some material from an earlier paper on the Transitional Support Program (Lange et al. 2003).

This work is made possible through a National Science Foundation grant (SBE-0123442). Any opinion, finding, and conclusion or recommendations expressed in this material are those of the author(s) and do not necessarily reflect the views of the National Science Foundation.

The authors are especially indebted to the late Dr. Denice Denton, the original principal investigator for the UW ADVANCE grant; and Mr. David Atsales, the former CIC program operations specialist, for handling fiscal administration of the grants. Members of the UW leadership team were very helpful and provided insightful advice about the structure of the program.

REFERENCES

American Association of University Professors. 2001. Statement of principles on family responsibilities and academic work. http://www.aaup.org/Issues/FamilyWork/Policy/policy.htm.

American Council on Education. 2005. An agenda for excellence: Creating flexibility in tenure-track faculty careers [executive summary]. http://www.acenet.edu/bookstore/pdf/2005_tenure_flex_summary.pdf.

Armenti, Carmen. 2004. May babies and post-tenure babies: Maternal decisions of women professors. *Review of Higher Education* 27:211–31.

Bailyn, Lotte. 1993. *Breaking the mold: Women, men, and time in the new corporate world.* New York: Free Press.

Barton, Leslie L., Alan D. Friedman, and C. J. Locke. 1995. Stress in pediatric faculty: Results of a national survey. *Archives of Pediatrics and Adolescent Medicine* 149 (7): 751–57.

Bertz, Elaine. 2003. Hidden disability and an academic career. *Academe-Bulletin of the AAUP* 89 (4): 51–53.

Blackburn, Robert T., and Carol Hollenshead. 1999. *University of Michigan faculty work-life study report.* Ann Arbor: Center for the Study of Higher and Postsecondary Education, University of Michigan.

Davis, D. E., and Helen S. Astin. 1990. Life cycle, career patterns, and gender stratification in academe: Breaking the myths and exposing truths. In *Storming the tower: Women in the academic world,* ed. S. S. Lie and V. E. O'Leary, 89–107. New York: Nichols/GP Publishing.

Gatta, Mary L., and Patricia A. Roos. 2002. *Balancing without a net in academia: Integrating family and work lives.* New Brunswick, NJ: Center for Women and Work, Rutgers University.

Gmelch, W. H., P. K. Willse, and N. P. Lourich. 1986. Dimensions of stress among university faculty: Factor analytic results from a national study. *Research in Higher Education* 24:266–85.

Grant, Linda, I. Kennelly, and K. B. Ward. 2000. Revisiting the gender, marriage, and parenthood puzzle in scientific careers. *Women's Studies Quarterly* 28:62–85.

Hensel, Nancy. 1991. *Realizing gender equality in higher education: The need to integrate work/family issues.* ASHE-ERIC Higher Education Report No. 2. Washington, DC: School of Education and Human Development, George Washington University.

Jacobs, Jerry A. 2004. The faculty time divide. *Sociological Forum* 19 (1): 3–27.

Jacobs, Jerry A., and Sarah E. Winslow. 2004. Overworked faculty: Job stresses and family demands. *Annals of the AAPSS* 596104–29.

Lange, Sheila, Eve A. Riskin, Suzanne G. Brainard, and Denice D. Denton. 2003. Implementing a transitional support program. Proceedings of the WEPAN 2003 Conference, Chicago, Illinois.

Mason, Mary Ann, and Marc Goulden. 2002. Do babies matter? The effect of family formation on the lifelong careers of academic men and women. *Academe* 88 (6): 21–27.

Sorcinelli, Mary Deane, and Janet P. Near. 1989. Relations between work and life away from work among university faculty. *Journal of Higher Education* 60 (1): 59–81.

Thompson, Carolyn J., and Eric L. Dey. 1998. Pushed to the margins: Sources of stress for African American college and university faculty. *Journal of Higher Education* 69 (3): 324–46.

Valian, Virginia. 1999. *Why so slow? The advancement of women.* Cambridge: MIT Press.

Varner, Amy. 2000. The consequences and costs of delaying attempted childbirth for women faculty. http://lsir.la.psu.edu/workfam/delaykids.pdf.

Williams, Joan. 2000a. *Unbending gender: Why work and family conflict and what to do about it.* Oxford: University Press.

———. 2000b. What stymies women's careers? It's personal. *Chronicle of Higher Education,* December 15, B10.

Wilson, Robin. 2001. For women with tenure and families, moving up the ranks is challenging. *Chronicle of Higher Education,* November 9, A11.

 PART THREE

Transforming Institutional Practices

NINE

Faculty Recruitment

MOBILIZING SCIENCE AND ENGINEERING FACULTY

Abigail J. Stewart, Janet E. Malley, & Danielle LaVaque-Manty

WHEN THE National Science Foundation (NSF) announced the Institutional Transformation grant program, it stated that "women scientists and engineers continue to be significantly underrepresented in some science and engineering fields and proportionately under-advanced in science and engineering in general in the Nation's colleges and universities. There is increasing recognition that the lack of women's full participation at the senior level of academe is often a systemic consequence of academic culture" (National Science Foundation 2001). The intervention outlined in this chapter aimed to address institutional factors that impede women's progress through the first doorway to a faculty career: recruitment into a full-time, tenure-track position. We believe that it is crucial to establish a "critical mass" of women faculty in every science and engineering department at the University of Michigan in order to reduce the salience of gender schemas—hypotheses about what men and women are like—that disadvantage women at all stages of their faculty careers. We use the term *recruitment* deliberately throughout the chapter to signify that hiring women faculty is necessarily a proactive process that often requires search committees to look beyond their usual applicant pools to find excellent women candidates.

Addressing recruitment raises broader questions about the evaluation of faculty members (in annual reviews, promotion and tenure reviews, and other settings), as well as everyday work environments for women. The specific approach we describe, though, focused narrowly and intensively on issues associated with recruitment per se. We assumed throughout that the larger work environment inevitably affected recruitment, and that the long-term impact of recruitment would depend on changes in the work environment. We also understood that recruitment practices alone could not effect significant long-term change. This systemic perspective was articulated in all activities outlined here, but we nevertheless maintained a strategic emphasis on recruitment, believing that doing so would not only have concrete results in the short term—increasing the number of women faculty members—but would also provide an opportunity to educate faculty about evaluation bias and climate problems that disadvantage women at every stage of their academic careers. Our efforts were no doubt enhanced and impeded by the relative success and failure of *other* efforts to improve *other* domains of institutional practice.

Our approach to hiring began from two empirically well-supported assumptions. First, women have been receiving an increasing proportion of the doctorates in science and engineering fields but do not apply for open faculty positions in numbers proportional to their doctorates. This suggests that women "choose" not to pursue tenure-track faculty positions (Sears 2003; Xie and Shauman 2003; Sonnert and Holton 1996); their choices may result from perceptions of their chances of academic success or of the costs of pursuing academic careers. Second, women on the tenure track advance more slowly and are likelier to exit than men (Valian 1999; Etzkowitz, Kemelgor, and Uzzi 2000; Long 2001).

Creating STRIDE to Increase the Number of Women in Science and Engineering

Given the gap between doctoral production and applications for faculty positions, waiting for women to apply for open positions will not increase the numbers of women faculty; a proactive approach is essential. Moreover, there is clear evidence that the climate must also change if recruited women are to stay and thrive (Fried et al. 1996; Stewart, Stubbs, and Malley 2002). At the UM, we created a faculty committee called Strategies and Tactics for Recruiting to Improve Diversity and Excellence (STRIDE) to improve recruitment and hiring of women scientists through peer education. Key tasks for the committee were increasing faculty awareness of issues involved in recruiting women, and providing conceptual and practical support to faculty eager to work on recruitment.[1]

It is not obvious to all faculty members that proactive recruitment of women is warranted, or even that it is not in itself a form of discrimination against men (Fried et al. 1996). Moreover, hiring is perceived by most faculty members as a key activity ensuring the vitality, excellence, and status of their intellectual community. Thus, it is crucial that STRIDE engage in open, respectful dialogue with faculty, admitting of a complex range of goals in any hiring process.

It is optimal if that dialogue takes place among individuals recognized by all as legitimate and equal stakeholders in the hiring process. For example, we suspected that scientists would be more receptive to ideas they might otherwise dismiss as unnecessary or "political" if they learned about them through colleagues they respected as researchers and individuals. Thus, STRIDE members are full professors in science and engineering fields who were nominated by the deans of their colleges to serve on the committee because the deans saw them as being both highly credible with other faculty and concerned about issues of diversity in science.[2] It should be noted, though, that none of them was associated with issues affecting women in science at the time that they joined the committee. Because women may be seen as "partial" to their own cause when addressing problems confronting women, more than half of the committee members are men.[3] The principal investigator for ADVANCE at the UM, who chaired the STRIDE committee, did not know the faculty members prior to inviting them to serve, nor did they know one another. Although STRIDE members volunteered time and expertise well in excess of what could be offered to them in compensation, all were provided with course release or research support in recognition of their work on behalf of the university community.

Development of STRIDE's Approach

In setting up the STRIDE committee, we learned a great deal from Harvard University's Committee on Faculty Diversity.[4] The UM ADVANCE project PI met with members of Harvard's committee before bringing the founding members of STRIDE together for the first time. She presented the Harvard committee's recommendations: acquire sufficient staff support for committee activities; engage in regular communication with deans; have multiple members meet with search committees; include both men and women in every meeting if possible; and be ready, with data, to talk about the demographics in each field. All of these became routine STRIDE practices. Other aspects of the Harvard committee's approach (e.g., a focus on informal meetings outside of official settings) seemed better adapted to Harvard's local culture than to Michigan's. We believe that other institutions will be able to adapt STRIDE's practices to suit their local conditions, just as we adapted those we learned from Harvard's committee.

Combining science and social science proved central to STRIDE's approach. The PI for ADVANCE at the UM is a professor in psychology and women's studies, familiar with social science research on psychology and sociology of gender, but not expert on the world of science and engineering. STRIDE members were drawn from three different colleges: the Medical School, College of Engineering, and College of Literature, Science and Arts. Experts in the subjects and cultures of their respective fields, they were largely unfamiliar with social science literature on gender. Over time, through intensive self-study and discussion, these scientists and engineers became experts on social scientific understandings of gender dynamics and cognitive bias and advocates of social science research on these topics as a source of insight into recruitment and hiring processes.

Self-Study and Teaching Others

The eight members of STRIDE first met in April 2002, having read several articles suggested by the PI about science and gender, and part of Virginia Valian's book *Why So Slow? The Advancement of Women.*[5] They watched a videotaped lecture by Valian and began to discuss the central problems involved in hiring and retaining women, both in their own experience and as examined in the social science literature. As they developed some consensus on the nature of the problem, and ideas about solutions, they also discussed the best means of communicating this information to their colleagues.

While the group found many of the articles they read useful (and were moved by the high level of replication and validation of basic findings in many studies), they were particularly impressed with Valian's synthesis of empirical research. They found the experimental results Valian reports in her book and lecture persuasive and believed their colleagues would, too. They began to build a presentation around key concepts presented by Valian and others. The central ideas that seemed crucial to convey were *critical mass, gender schemas, evaluation bias,* and *accumulation of disadvantage.* They wanted to stress that (1) men and women *both* rely on gender schemas, or hypotheses about what men and women are like, (2) these schemas lead them to underevaluate women in science and engineering and other professional settings, and (3) small disadvantages accrue to women in ways that build over time. Women scientists are often told not to make mountains out of molehills, but as Valian notes in her lecture, mountains *are* molehills piled one on top of the other. Moreover, when women are underrepresented, their gender is particularly salient, triggering reliance on gender schemas.

These concepts were the building blocks of STRIDE's understanding. The next step was to apply them to academic recruitment, ideally by tying them directly to potential recruitment practices. STRIDE sought to identify strategies that would (for example) lower the salience of gender, minimize reliance on gender schemas, and help committees fairly evaluate candidates. STRIDE members found their discussions of potential recruitment strategies and approaches to persuading colleagues of these points intellectually invigorating and deeply mobilizing. Deciding that their job was to meet with anyone with input into hiring, including department chairs, search committees, and entire departmental faculties, they worked with the ADVANCE project's program manager,[6] who provided staff support, to design the materials they would offer prospective audiences. This entailed meeting for more than twenty hours over the course of four months, with reading and email exchanges between sessions.

STRIDE designed a PowerPoint presentation that would serve as a road map for discussions with various audiences. Though individual committee members could (and did) alter it as needed, the goal was to create a fairly uniform framework for STRIDE presentations. They designed and discarded many slides while developing consensus about what data were most useful and which intellectual points most central. Eventually, they configured the presentation so that the level of detail offered could be tailored to the interests of particular audiences through pop-ups and links. They also collaborated in writing a twenty-seven-page recruitment handbook.[7] Arguing over the specific content of these materials allowed committee members to air concerns and differences and create common priorities, forging a collective understanding that underlies their ongoing mobilization as "activists" on behalf of gender equity in academic science and engineering. For example, they argued at length over whether or not it would be helpful for search committees to let women candidates know that they were trying hard to hire women. The men thought it would send a positive signal that the department was attentive to gender issues, while the women thought it would make the candidate feel that she was being considered for reasons other than the quality of her research and teaching. In the end, the men came around to the women's point of view, and STRIDE agreed to advise departments to "recruit women scientists as *scientists,* not as women."

Between September 2002 and April 2003 STRIDE made twenty-six presentations and distributed over three hundred copies of the handbook. Most presentations were made to entire departmental faculties or search committees, but audiences also included deans and other groups of administrators. In addition, after they met with the dean of the College of Literature, Science and Arts, he adopted the handbook as part of the College's recruiting process and distributed it to all College search committees, even those that did not invite STRIDE to give presentations. Finally, two dean search committees consulted with the STRIDE committee chair about their hiring processes and also received the handbook.

Changing STRIDE Activities over Time

Development of FASTER. STRIDE's activities have evolved since that first year. During the summer of 2003 the committee created a program to develop additional colleagues who would have a fuller understanding of the issues and who might eventually become new members of STRIDE. They designed a six-hour program (which took place in two half-days), based on their own past curriculum, to share what they had learned with selected colleagues. STRIDE hoped to replicate its experiences of the previous year and cultivate an expanded group of "activists." Because having specific tasks and problems to solve (e.g., a presentation to design and a handbook to write) had given them concrete ways to apply new theoretical insights, they gave their colleagues something to work on as well: helping to improve STRIDE's approach to departments and search committees. The new group named itself Friends and Allies of Science and Technology Equity in Recruiting (FASTER). STRIDE has since acquired two new and enthusiastic members, from the pool of allies who attended FASTER sessions.

New Goal Setting. At the end of its first year, STRIDE decided it was important to address some issues better and requested material for continuing self-education. Concerning them most were the impact of family and personal life issues on women scientists and engineers, pressure to spend long hours in the lab or office, and the importance of race and ethnicity as well as gender. They studied these concerns intensively and added material to their presentation to cover them.

In addition, the committee believed that its presentations were not interactive enough. Being scheduled into regular department meetings (which had seemed desirable at first) made time short and left people distracted by other agendas. They believed that their own insecurities discussing the issues also reduced their effectiveness. In the beginning they preferred an easier-to-control lecture mode over an interactive seminar mode. Now, though, they were ready for more interaction. They redesigned the presentation to include less data, fewer didactic points, and more discussion.

Collaborating with CRLT. During the next year, fewer departments, committees, and groups requested STRIDE's presentations, partly because of budget constraints on hiring. Even where there were searches, chairs felt they had already brought STRIDE in the year before, and department colleagues were unsure whether further interaction with STRIDE would be useful. However, STRIDE was newly engaging in other activities. For example, as a way of making a commitment to facilitate women's success and retention, the committee began introducing itself to incoming women faculty who might turn to them for advice in the future. It also developed a relationship with another group working on climate issues through the ADVANCE Project, the Center for Research on Learning and Teaching (CRLT) Players.

CRLT Players, an interactive theater group that specializes in depicting teacher-student interactions, was commissioned to develop sketches for ADVANCE. The first sketch, portraying a faculty meeting in an engineering department, was designed to reveal problematic dynamics where women are a small minority of the faculty. The topic of the "meeting" involves a hiring decision where one candidate is a woman. The sketch is too complex to summarize here and is discussed in detail elsewhere in this volume (chapter 13), but it provides rich material for discussion. CRLT has highly skilled facilitators, but they depend on audiences to ask good questions to get discussion started. STRIDE realized that it could play a role in those discussions by sending some members to attend performances to ask key questions audiences might not come up with otherwise.

Finally, the provost asked the ADVANCE PI to make a presentation to every search committee for new deans, identifying gender-equitable practices searches for high-level administrators could employ. She distributed STRIDE's recruitment handbook and presented STRIDE-based "talking points" at these meetings.

Administrative Support. By the end of the second academic year, it seemed clear that STRIDE's original process of engagement—through department meetings—was restricting its contributions to campus hiring. The ADVANCE Steering Committee, which included the deans of the Medical School and the colleges of Engineering and Literature, Science and the Arts, invited STRIDE to develop a workshop that could be delivered to larger, more heterogeneous audiences of faculty across the three colleges. They proposed that they could mandate that all search committee chairs in their three colleges participate in STRIDE workshops focused on recruitment. In addition, two of the deans instituted a form of "short-list review," whereby slates of candidates for on-campus interviews (the "short list") were compared with the pool of applicants. If a demographic group was less well represented in the short list than in the pool, the dean's office had an opportunity for a conversation with the department or search committee chair. Equally, if a group was underrepresented on both lists, this could be discussed. The short-list review offered, then, a formal opportunity for discussion of the degree to which the search successfully identified diverse candidates, and if it had not, why.

In response to the deans' interest, STRIDE designed a much longer structure for interaction with search committee chairs, and offered three two-and-a-half-hour workshops during fall 2004. These included brief presentations on six separate topics: (1) what is the problem?, (2) why diversity matters, (3) unconscious bias in evaluation, (4) recruitment strategies, (5) dual-career and family policies, and (6) how family matters for evaluation bias. The PI introduced the presentations, summarized key points at the end, and facilitated discussion of issues raised. This format offered fuller coverage of each issue. A total of fifty-nine faculty heading or serving on search committees in the three colleges participated in one of the three workshops. Both STRIDE and the participants found that extended discussion, with full coverage of many topics, was much more productive than briefer interactions. In addition, the search committee chairs were highly motivated by their roles to focus on what STRIDE had to offer. And STRIDE worked hard to help search committee chairs, pointing to very specific practices and providing new materials for their use. For example, they developed a new tool for rating job candidates that would help minimize reliance on stereotypes by focusing judges on specific and individuating evidence.[8]

STRIDE's Future

No doubt STRIDE will continue to evolve as institutional demands change. It is currently broadening its mandate, at the provost's request, beyond science and engineering. Over time, the University plans to support the committee centrally (without NSF funds). As STRIDE moves toward that model, it must expand its focus to include all fields, incorporate new members, and eventually replace the current group entirely. It is hoped that FASTER will continue to serve as a conduit for recruiting new participants in STRIDE, as it did with its first pair of new recruits. This is the outgrowth of its own success and of an active partnership with the deans on the steering committee and the provost. Meanwhile, it is fair to ask whether there is evidence that STRIDE has been effective. While we cannot claim that STRIDE alone has had any particular effects, we have gathered data that suggest that it has had significant impact. Our data include evaluations by audiences of STRIDE presentations, actual recruitment and hiring of women during the first two years of STRIDE's operation, and reported effects of participating on STRIDE members themselves.

Measuring STRIDE's Impact

Reception of STRIDE by Faculty Members

ADVANCE project staff conducted two web surveys of faculty members: the first in departments that hosted STRIDE presentations in the first year; the second of workshop attendees in fall 2004. Administrators and other nonfaculty or nondepartmental groups were not included in the survey.

Survey after the first year. The first web survey was sent to all faculty in relevant departments, via e-mail using department e-mail lists. Twenty-eight faculty (about 20% of those surveyed) who had attended presentations completed surveys. Because the surveys were anonymous, we were unable to attach individual responses to departments or the specific presentation each respondent attended. The survey asked respondents to rate the effectiveness of the presentation(s) on a scale from 1, "not at all effective," to 5, "very effective." Open-ended questions asked what was most and least effective about the presentation(s) and what, if any, effect the presentation(s) had on their respective departments and their search processes.

Overall ratings ranged from 2 to 5, with a skew toward favorable ratings (seventeen of twenty-eight giving the top two ratings). A quarter gave a negative rating of 2 (seven of twenty-eight). This is not surprising, since the most motivated individuals—those who very much liked or disliked the presentation—were likely to respond to a web survey. The presentation received more positive ratings than negative ones (over 60% rated the presentation as "very effective" or "effective"), and nobody rated the presentation "not at all effective." One respondent explained that the presentation was "excellent," but rated it "not very effective" (2) because the faculty member thought it had no impact on this individual's department's searches.

Open-ended comments about the most and least effective aspects of the presentations provided further context for these ratings. The paramount value of the presentations to several respondents was that senior faculty were taking the time to bring these issues out into the open for discussion, lending credibility to their importance.

Respondents also noted the quality of the presentations, describing them as "excellent," "well-argued," "consistent," and "professional." Two issues appeared to be reflected here: STRIDE's thorough discussions of both *demographic data* about women's training and hiring history and *gender bias,* including a clear review of research in this area. Respondents were also impressed by STRIDE's summary of research on gender bias and schemas, particularly the examples illustrating how well-intentioned behaviors can unwittingly result in bias. STRIDE relied on video clips from a talk by Virginia Valian (2001) that several mentioned as particularly effective in making this point.

Some respondents expressed concerns that the presentations were not effective in reaching some faculty, variously pinpointing men, chairs, or senior faculty. A few respondents were not themselves persuaded by the information STRIDE presented. One reported, "I remain unconvinced by the main hypothesis." Some reported that they "felt accused of being sexists." On the other hand, some felt STRIDE's message was watered down so it would not be offensive, and as a result, was less effective. One respondent was optimistic that the positive impact of STRIDE's presentations might not be immediate. "There was certainly some reaction against the presentation . . . but I believe there was also some very useful information that my colleagues will have reflected on, as I have."

Many respondents felt the STRIDE presentations had a positive effect on their departments, most often by making faculty more aware of the issues. In all, eleven of the twenty-seven faculty who responded to this question (41%) indicated that the STRIDE presentation had had some positive effect on their departments and how they conduct searches. In contrast, thirteen (48%) indicated that there was no clear change in department hiring practices as a result; the remaining three respondents (11%) didn't know if STRIDE had had an effect or not. Thus, there was evidence from the first survey that STRIDE had a positive impact on the recruitment process. There was also evidence that change was not uniform; STRIDE was able to use the critical feedback in designing the workshops for fall 2004.

Reception of STRIDE Workshops

After holding the newly designed workshops, UM ADVANCE's evaluation staff sent an on-line survey to the fifty-nine attendees. Twenty-three of the twenty-six respondents (88%) rated their workshop overall as very effective or somewhat effective; three attendees gave a neutral rating. There was relatively little variation in specific topic ratings, though the section on "unconscious bias in evaluation" received the most uniformly positive rating. No respondents reported a "not at all effective" rating for any topic. Attendees were also asked what was most effective. The three most common responses to this question were (1) the presentation was well supported by data and substantive research; (2) the workshop provided specific, practicable strategies and recommendations; (3) the presenters were focused, enthusiastic, and knowledgeable.

Regarding what was least effective, some respondents said presenters spent too much time trying to convince the audience that the problem exists and not enough time on details and strategies, and that they found the style of presentation or the workshop atmosphere flawed. Respondents offered a variety of suggestions for improvement, mainly focusing on making the workshop more interactive, and focusing even more on solutions and strategies.

With respect to how the workshop might improve searches in their departments, most respondents focused on its effects on their own roles in the search process. One commented, "My own reading of letters has been altered by the impact of the data in the paper [we read]." Another asserted that "if . . . very few women are in the pool, we will . . . solicit more names and applications." Another wrote, "[The workshop] has already affected [the respondent's role] (e.g. by my pushing a colleague to tell me the outstanding *women* in addition to his short-list of outstanding candidates)." One search chair commented, "I will strongly recommend that all members of the committee have written notes on each candidate, because the workshops helped me realize that this is probably the most common way that bias can enter the process. I will also encourage my committee to be able to justify why they are removing applications from the acceptable pile."

A clear majority of respondents supported the idea of offering this kind of workshop on an annual basis. A few recommended expanding the target audience to include all faculty (particularly new department chairs and directors) as well as administrative staff who support search committees. One commented, "It would be a serious mistake to have this kind of training program developed and *not* used to reach everyone and keep the momentum going."

The longer workshop format, targeting search committee chairs, was clearly better received than earlier presentations to more diverse faculty groups. It is possible that more sympathetic individuals were asked to head searches, and therefore the audience for the workshops reflected a narrower band of opinion. However, these were precisely the faculty most able to influence recruitment through their roles as search committee chairs.

Impact on Hiring

Another aspect of STRIDE's effectiveness can be measured by comparing the proportion of women scientists and engineers hired during the most recent academic years to the proportion hired in the past. Table 1 shows the proportion of men and women hired in each of the three colleges that employ the largest number of scientists and engineers at the University of Michigan over the last four academic years. Note the marked, and statistically significant, increase in the proportion of women hired comparing the two pre-STRIDE years (AY2001 and AY2002) with the three post-STRIDE years (AY2003, AY2004, and AY2005) (chi-square = 10.33, p = .001).

During the academic year (AY) 2003 recruitment period, the absolute number of women hired in the three colleges employing the largest number of scientists and engineers increased between three- and

TABLE 1. Men and Women Hired in Natural Science and Engineering Departments in Three University of Michigan Colleges

College	AY2001		AY2002		AY2003		AY2004		AY2005	
	Men	Women	Men	Women	Men	Women	Men	Women	Men	Women
Medical School (Basic Sciences)	1	1	1	1	6	2	7	5	5	2
College of Engineering	25	2	8	1	17	8	6	4	13	5
LSA (Natural Science Division)	15	3	13	2	9	9	6	3	19	8
Total % Women		13%		15%		37%		39%		29%

fourfold (from six and four to nineteen). While University-wide bud-getary constraints resulted in a drop in total number of faculty hired in AY2004, the improved pattern persisted; the proportion of women hired in those colleges more than doubled in AY2003 and AY2004 when compared to the two preceding years.

While many factors no doubt contributed to departments' willing-ness to hire more women, STRIDE is the intervention that most directly provided tools and ideas to aid in recruitment. Moreover, of six-teen women hired in the first year, twelve were hired into departments that had STRIDE presentations (and all chairs were exposed to STRIDE presentations and could have made use of the recruitment handbooks). It is reasonable to conclude that STRIDE, having addressed so many audiences and drawn specific attention to useful poli-cies and resources, contributed to the increase in the number of women hired. For example, many search committees and department members STRIDE met with were unaware that the University had resources to aid in placing spouses and partners of new faculty, and some were unaware of University policies regarding maternity and the tenure clock. Further, anecdotal evidence suggests that at least one department successfully recruited a highly regarded female candidate away from strong competitors partly because of specific advice from STRIDE.

Impact on Committee Members

Seven of the eight founding members of STRIDE were interviewed in fall 2002, shortly after they began making public presentations. They were interviewed again in December 2003, when they had significantly more experience working together and interacting with departments and search committees. (The eighth original member of STRIDE retired from the UM in 2002.)

In the first round of interviews, all seven STRIDE members indi-cated that their summer "study sessions" had a strong impact on their understanding of the problems women science and engineering faculty confront. They were impressed by what social science could tell them. One of the men on the committee put it, "I was surprised by the num-ber of studies . . . on the nature of the bias, and where the bias comes from . . . there's been a lot of really, really good research that's been done . . . these studies are fantastic."

The women found themselves rethinking not only the gendered

nature of science and engineering, but also the way their own careers have been shaped by dynamics that they may have chosen, in self-defense, not to think about: "I would say the most surprising thing that I've learned was what unconscious bias was, and how prevalent it is, and how it works. . . . I discovered . . . that in my own career I had been coping by denying that I had ever had any problems . . . it turned out to be much more of an emotional voyage than I had ever expected it to be."

In the interviews in December 2003, STRIDE members referred back to their "study sessions," again describing their summer 2002 conversations as part of a period of discovery, or, as one member put it, "consciousness-raising." Another member said:

> The process that we went through worked so well . . . the process being to identify a group of senior faculty, both men and women, in the sciences and engineering, who have shown some evidence of being concerned about these issues, but who clearly don't know all the literature. And I think every one of us . . . is like that. . . . We all had some previous commitment, but what we realized when we got together and started actually looking at the data and learning together was that we didn't understand, we didn't really know what was going on, we really were quite naïve . . . that discovery process, I think, was so critical to building the passion that the current group has.

More than a year elapsed between the first set of interviews and the second, while the STRIDE tem continued to read articles and studies on gender and science and to give presentations throughout the university, gaining increasing confidence in their ability to present this information persuasively. All seven STRIDE members reported in the second interview that they were as motivated as before, and felt more assured in their roles, both in giving slide presentations and in dealing with colleagues informally. Further, all mentioned intervening in negative gender dynamics in their departments and other settings in ways they believed they would not have prior to experience with STRIDE.

One man said, "You could say, perhaps, that I've become a bit more combative; things that I used to just shy away from, I now feel that . . . I'm obligated to do something about." Similarly, one woman reported,

"I think that I am much more willing to discuss these issues with my colleagues and much more willing to intervene in cases where I feel that this [gender bias] is an issue, and to intervene in terms of saying, ok, I think we need to look at x, y, and z, and the literature says these things, and I just want us to consider that . . ." Another woman said, "I use that knowledge to enrich my discussions with my colleagues in other contexts, when I'm involved in making decisions on committees . . . or at faculty meetings. . . . I find myself able to draw on a much larger wealth of information about gender-related issues . . . that I think is directly a result of being a member of STRIDE and learning what we did in that process." She reported successfully encouraging department chairs in her college to take advantage of other ADVANCE initiatives, such as the Departmental Transformation Grants. "That's an example of something I would not previously have had the motivation, nor the . . . kind of a sense of responsibility, I guess, that allowed me to say [what I said] to these chairs."

One man noted the importance of having his fellow STRIDE members as an alternative set of peers outside his department: "One of the big values in STRIDE for me has been just knowing that other people feel the same way, and they're struggling with the same issues. When you're isolated, as you would be in a department, it isn't normal to talk about these things."

One question STRIDE members were asked in 2002 was whether they could sustain their current pace for the remaining four years of the ADVANCE project. More than one suggested that the long time horizon enabled them to think more expansively about what was possible and believe there was enough time to accomplish something.

As noted above, STRIDE members not only continue to be committed to their collective project of educating search committees about evaluation bias and recruitment strategies, they have also become more inclined to speak out about gender issues on their own in other contexts. One found himself advising colleagues at other universities about applying for ADVANCE grants. Another reported that she is now much more willing to mentor other women with respect to gender issues. Another made women in science (in his own field, in particular) the subject of a public lecture associated with an award he won. All described a consistent level of activism motivated by their greater empirical knowledge of problems confronting women in science and tools for addressing them.

Conclusions and Recommendations

STRIDE's work is still in progress. However, its early success in improving recruitment and hiring of women science and engineering faculty at the UM leads us to recommend it as a model that might be adopted by other institutions. We believe that the long-term effects of mobilizing these highly respected scientists can only be positive for women faculty. STRIDE constitutes a core group of advocates, not only well intentioned and strongly motivated, but also unusually well informed, combining expertise in science and engineering with an understanding of social science literature on gender. We incorporate here both our own observations about factors that may have made an important difference in STRIDE's formation and practice, and those offered by members of the STRIDE committee.

1. The request to serve on STRIDE came from a campus-wide project thoroughly legitimated by the central administration and associated with an institutional commitment to a long-term process of change.

2. Concrete resources were provided to compensate for time spent on the committee. This not only provided enabling conditions to faculty members, but also communicated the seriousness of the institutional commitment.

3. The commitment to serve extended over several years. This persuaded many committee members that the duration of the planned effort matched that required to create and sustain change.

4. The committee felt supported in its work by having access to an "expert" on the social science literature on gender and a support staff competent to help members find the empirical and theoretical literature they were interested in and help them implement their ideas.

5. The committee worked together over an extended period defining and redefining its own message and strategy; its autonomy and capacity to define its own "charge" were, and remain, highly valued.

6. The committee included participants from each of the most significant relevant environments (the three large colleges), which gave it lots of relevant experience and examples to analyze once it had tools to do so.

7. The committee included both men and women, all highly respected in their fields.

8. The committee thrashed out hard issues, building trust in each other and confidence in their message.

9. The committee processed experiences on a frequent basis, using feedback to modify its activities. A sense of growth made it feel effective in responding to changing conditions.

STRIDE members are the first to admit that it is not possible to persuade everyone to care about gender equity or believe that an equitable system and culture are not already in place. Indeed, they convey to colleagues both the very slow pace of change and the consequent need for long-term commitment and persistence. They believe, though, that starting a conversation about the issues involved is a useful first step. We believe that the STRIDE approach is one promising way to start that conversation and set the stage for an ongoing process of transformation of institutional recruitment practices.

NOTES

This chapter draws some material from an earlier paper on the STRIDE committee (Stewart, LaVaque-Manty, and Malley 2004).

The authors are particularly grateful to the founding members of STRIDE for their amazing dedication and service, as well as for their advice on this manuscript: Anthony England, Carol Fierke, Melvin Hochster, Sam Mukasa, Martha Pollack, Pamela Raymond, Michael Savageau, and John Vandermeer. The "new" additions to the committee, Gary Huffnagle and Wayne Jones, have certainly equaled their standard, and we are grateful to them too. We are also grateful to Deans Stephen Director, Allen Lichter, Terrence McDonald, Ronald Gibala, Janet Weiss, and David Munson for their support to STRIDE; and to President Mary Sue Coleman, Provost Teresa Sullivan, and Associate Provost Lori Pierce for their support to, and recognition of, STRIDE's contributions. STRIDE would not exist if it were not for the NSF ADVANCE Institutional Transformation program, and its director, Alice Hogan. We are grateful to both for the visionary leadership reflected in the program and for the financial support and practical advice they have provided.

1. The acronym originally stood for Science and Technology Recruiting to Improve Diversity and Excellence but has since been revised. See Stewart, LaVaque-Manty, and Malley 2004 for a more detailed study of STRIDE's early development.

2. The initial committee included Pamela Raymond and Michael Savageau from the School of Medicine; Anthony England and Martha Pollack from the College of Engineering; and Carol Fierke, Melvin Hochster, Samuel Mukasa, and John Vandermeer from the College of Literature, Science and the Arts. Michael Sav-

ageau retired from UM and the committee at the end of the first year. Gary Huffnagle from the School of Medicine and Wayne Jones from the College of Engineering joined the committee in December 2003.

3. Five of the original eight; six of the current nine.

4. See http://schwinger.harvard.edu/~georgi/women/cfd.html for information about Harvard's committee.

5. See http://www.umich.edu/~advproj/stride.html for further information
about these materials. STRIDE's presentation road map, as well as the handbook,
are available at http://www.umich.edu/~advproj/stridepresents_files/frame.htm.

6. Three individuals—Danielle LaVaque-Manty, Robin Stephenson, and Cynthia Hudgins—have provided outstanding support to STRIDE in this role.

7. Available on the project web site: http://www.umich.edu/~advproj/hand-
book.pdf.

8. Available on the project web site: http://www.umich.edu/~advproj/candi-
date_evaluation_sheet.pdf.

REFERENCES

Etzkowitz, H., C. Kemelgor, and B. Uzzi. 2000. *Athena unbound: The advancement
    of women in science and technology.* New York: Cambridge University Press.
Fried, L., C. Francomano, S. MacDonald, and M. Wagner. 1996. Career develop-
    ment for women in academic medicine: Multiple interventions in a department
    of medicine. *Journal of the American Medical Association* 276 (11): 898–905.
Long, J. Scott, ed. 2001. Executive summary. In *From scarcity to visibility: Gender dif-
    ferences in the careers of doctoral scientists and engineers.* Washington, DC: National
    Academy Press.
National Science Foundation. 2001. ADVANCE: Increasing the participation and
    advancement of women in academic science and engineering careers. Program
    announcement, nsf0169. http://www.nsf.gov/pubs/2001/nsf0169/nsf0169
    .htm.
Sears, A. W. 2003. Image problems deplete the number of women in academic
    applicant pools. *Journal of Women and Minorities in Science and Engineering* 9 (2):
    169–81.
Sonnert, G., and G. Holton. 1996. Career patterns of women and men in the sci-
    ences. *American Scientist* 84:63–71.
Stewart, A. J., D. LaVaque-Manty, and J. E. Malley. 2004. Recruiting women fac-
    ulty in science and engineering: Preliminary evaluation of one intervention
    model. *Journal of Women and Minorities in Science and Engineering* 10: 361–75.
Stewart, A., J. Stubbs, and J. Malley. 2002. Assessing the academic work environ-
    ment for women scientists and engineers. http://www.umich.edu/~advproj/
    climatereport.pdf
Valian, V. 1999. *Why so slow? The advancement of women.* Cambridge: MIT Press.
———. 2001. The advancement of women in science and engineering: Why so
    slow? Rice University webcast, March 29. http://www.rice.edu/webcast/
    speeches/20010329valian.html.
Xie, Y, and K. Shauman. 2003. *Women in science: Career processes and outcomes.* Cam-
    bridge: Harvard University Press.

TEN

Scaling the Wall

HELPING FEMALE FACULTY IN ECONOMICS
ACHIEVE TENURE

Rachel Croson & KimMarie McGoldrick

AS IN MANY sciences, female representation in economics has grown, but hurdles still exist. Perhaps the most difficult hurdle is the transition from a junior position to a senior one. In academia, this generally takes the form of promotion from assistant to associate professor with tenure. The tenure hurdle is difficult for all academics, but recent evidence suggests it is disproportionately difficult for women, and more difficult for women in economics than in other fields (Ginther and Kahn 2004). After increasing steadily from 1974 through 1990, growth in the representation of women at the associate professor level in economics has essentially halted during the past decade (Kahn 2002).

Based on this evidence, the Committee on the Status of Women in the Economics Profession of the American Economics Association undertook an intervention, a series of workshops, designed to help female junior faculty achieve promotion and tenure. The initiative was funded jointly by ADVANCE and the Economics Panel of the NSF. We believe that similar programs can be implemented both by other disciplines facing similar challenges and by institutions looking to design an intervention at this critical stage of professional development. In this chapter we describe our workshops, discuss some of the lessons we learned, and offer suggestions for amending our procedures for use by institutions.

Motivation and Background

The Committee on the Status of Women in the Economics Profession (CSWEP) was founded in 1971 by the American Economics Association (AEA) in order "to eliminate discrimination against women, and to redress the low representation of women, in the economics profession." CSWEP has worked in a variety of ways to fulfill this mandate. CSWEP fields an annual survey of women's representation as students and faculty members in economics departments, publishes a thrice-yearly newsletter with articles of particular interest to women, and arranges sessions and hosts social events at the annual meetings of the AEA and the regional economics associations. CSWEP sessions at the meetings of the AEA and regional associations have explored issues related to gender, as well as substantive issues, and have also provided a vehicle for junior women to get on the programs of these meetings by organizing sessions. CSWEP was also the moving force behind the establishment of a child care program at the annual meetings of the AEA.

While CSWEP has recorded and, in some cases, facilitated many gains, there remains a significant difficulty in increasing the representation of women in the ranks of senior (tenured) faculty. In 2001 CSWEP sent a questionnaire to representatives at 115 Ph.D.-granting institutions in the United States and found that their staffs include 22% female assistant professors but only 16% female associate professors with tenure. Approximately 6% of the full professors were women, despite the fact that the participation of women at the lower ranks has held steady at around 20% for more than twelve years. The report can be found in the 2002 issue of the *AER Papers and Proceedings,* or at www.cswep.org (Committee on the Status of Women in the Economics Profession 2002).

Using the methodology described by Blank (1993), the report calculates that if decisions to advance assistants to the associate rank were equally likely for men and for women, the expected proportion of female associate professors would be 20% rather than 16%. The report concludes that "women are not advancing to the Associate Professor rank" (2002, 519) and that if they were, "the share of Associate Professor positions would have been between 5% and 15% higher than it was in 2001" (2002, 518–19). A recent study by Kahn (2002) comes to similar conclusions. While substantial increases in the percentage of women among associate professors were expected throughout the nineties (based on growth in the female percentage of assistant professors in earlier years), these increases did not materialize.

These numbers are also disappointing when compared with other fields. According to the U.S. Department of Education, 41.4% of faculty in degree-granting institutions are female compared to 13.1% in the CSWEP database (Zimbler 2001). Not only does economics lag behind other academic fields in female representation overall, the drop-off exhibited between the proportion of female untenured and tenured faculty that we see in economics does not exist in other fields. Results of a survey of the top fifty chemistry departments found a negligible drop-off of the proportion of women between the assistant and associate professor ranks—20.7% women at the assistant professor level and 20.5% at the associate professor level (Nelson 2005). A study by Ginther and Kahn (2004) suggests that the slower progress of women in economics as compared with other fields is not explained by their relative qualifications. They find that, controlling for measured characteristics, gender differences in promotion rates are considerably larger in economics than in both the humanities and the natural sciences.

This and related evidence suggested to us that women in economics have a particularly difficult time making the transition from untenured to tenured professor, compared both with men in economics and with women in other fields. We decided that an intervention was needed in order to aid in this transition and bring economics to parity with other sciences in this area. To decide what we should do, we turned to the literature from our own field of labor economics.

Our Approach

One concern that we discussed was the possibility that, on average, women simply didn't know as much about getting promoted as men. For example, results from McDowell, Singell, and Slater (2002) suggest that one possible cause of the problem of the leaky tenure pipeline for women may be a paucity of collegial networks. They find that female economists are less likely to coauthor than their male colleagues, even after controlling for publication rates. Blau, Ferber, and Winkler (2002) point out that, due to their minority status, women in predominantly male fields may not be included in the informal networks that form more readily among their male peers. Such groups can be helpful in sharing information and providing support and encouragement. We attempted to create these networks among women (junior-junior via peer networks and junior-senior via mentoring relationships). We also wanted to encourage women to create these networks with others at their institutions and in their disciplines by offering tools and tips for publicizing their work, networking at professional conferences, and developing relationships at their institutions.

Based on these considerations, we developed on an intervention aimed at the untenured (junior) women designed to help them achieve promotion and tenure. Research suggests that one form of formal assistance that may be effective is mentoring (see Ragins 1999 and Zachary 2000 for reviews of the literature). Blau, Ferber, and Winkler argue that mentoring serves three separate purposes. First, mentoring provides *role models* for young academics; the junior party learns by simply watching the behavior of the more senior. Blau et al. argue that the lack of senior women in predominantly male fields like economics puts junior women at a disadvantage in this arena. In particular, junior women "have inadequate information about acceptable (or successful) modes of behavior" (2002, 177).

Second, mentoring offers *information transmission:* senior faculty share tricks of the trade and strategies for success with junior faculty. Junior women who are undermentored are likely to "lack access to the knowledge that older women have acquired about successful [career] strategies" (2002, 177). This might include particular insights about combining work and family, details about journal editors' tastes and other professional insights, successful teaching strategies, and advice on interacting with one's administration or department chair.

Third, mentoring establishes *informal relationships*. Again, Blau et al. argue that junior women are often excluded from these relationships.

> The mentor-protégé system is generally the result of the older individual identifying with the younger person. Male mentors may simply not identify with young women. Or they may fear that the development of a close relationship with a young woman would be misunderstood by their colleagues or their wives. (2002, 177)

We thus sought an intervention that included these three features, *role models, information transmission,* and *informal relationships* between junior and senior academics. To these we added a fourth, that of *peer networks* based on the research of McDowell, Singell, and Slater (2002). (See also McDowell and Smith 1992; and McDowell and Melvin 1983).

We decided to hold a series of workshops, bringing together junior and senior faculty in economics from many universities. These experiences would expose junior faculty to role models they would otherwise not see, as in many economics departments there are no senior women at all. They would serve as devices for information transmission. Finally, we strove to create relationships between junior and senior academics in the same field, and to create peer networks among junior academics. Our vision was that the workshops would provide some information transmission, create relationships, and initiate the mentoring processes, but that the information transmission, informal relationships and networks would continue after the workshops had ended. Thus, more generally, our intervention addressed gender-specific issues like isolation of women and lesser access to networks as documented in the literature.

Implementing the Workshops

We held multiple workshops, designed to fulfill the needs of academics at different types of institutions. The national workshops would be held in conjunction with the AEA meetings in January, and would target junior faculty at primarily research-oriented institutions. The regional workshops would be held in conjunction with the regional economics meetings (Eastern Economic Association, Southern Economic Association, Western Economic Association, and Midwest Economic Association) and would target junior faculty at institutions where promotion is based on a combination of research, teaching, and service. This separation turned out to be important, as the advice offered (information transmission) is very different for the different target audiences, as is the choice of senior mentors. Furthermore, the focus of the workshops would be different, with the national workshops focusing almost exclusively on research, the regional workshops on a mixture of research and teaching.

As suggested above, workshops were designed to provide information transmission and role models, and develop informal relationships. In order to facilitate this process, small groups (between four and six persons) with common research and teaching interests would form the basis for interactions. As a result, the first criterion for accepting applications was in terms of goodness of fit in the construction of groups. Additionally, since the focus of these workshops was on helping women achieve tenure and promotion, only untenured faculty were considered for participation. In the case of the national workshops, twice as many applications were received as space allotted; thus, qualified participants were first grouped according to interests, and then approximately one-half were randomly accepted.

It is important to note that regional and national workshops targeted different clientele. Issues faced by women at Ph.D.-granting institutions are different from those faced at other institutions. These include differences in resources, expectations, and service loads, to name a few. We believed that separate workshops would provide the best mentoring experiences, as the sessions could better focus on the relevant issues for each group. For example, managing a teaching or research assistant is simply not often a concern for faculty at liberal arts institutions.

In order to stretch the budget as far as possible, organizers of national and regional workshops donated their time to develop and advertise the workshop, screen participants, and facilitate each workshop. Participants were expected to pay for their own travel expenses, although meals and housing were provided. (It is important to note that most institutions reimbursed travel expenses since the workshops were held in conjunction with national and regional economic association meetings.)

Identifying mentors was our second task. We created a list of tenured female faculty at institutions like those we were targeting for the workshops (research-focused universities for the national workshops, and liberal arts institutions for the regional workshops). The list was surprisingly short. We approached candidate mentors, explained the purpose of the workshop and the role we were asking of them, and invited them to participate as a mentor. We also stressed that their role would not end when the workshop ended, but that we hoped that relationships would form between participants and mentors, who would continue in a mentoring relationship in upcoming years.

We offered no honorarium, but did pay travel expenses, lodging, and meals for mentors. Because the workshops were held in conjunction with a regional or national conference, we had hoped that potential mentors would find the travel reimbursement an attractive incentive. In practice, however, most mentors agreed to participate because they believed in the mission of the workshops, rather than for financial gain. We assigned one senior mentor (full professor) and one junior mentor (recently tenured) to each group.

Workshops were developed to provide two full days of interactions.

(See table 1 for a detailed description of a sample workshop.) Panel plenary sessions facilitated the transmission of key information to all participants, while small team sessions allowed participants to interact closely with successful role models and obtain situation-specific advice. Although the two-day structure is not required for a workshop of this nature, we believed that it was necessary in order to cover the number of significant topics included, enhance bonding of groups, and increase the probability of their interactions beyond the workshop. Providing the opportunity to address these topics in a concentrated time period provided participants with a sense of immersion and made them more likely, as the workshop progressed, to openly discuss sensitive issues.

TABLE 1.   CSWEP National Mentoring Workshops, Sample Agenda

Introduction	
4:00pm–5:00pm	Registration
5:00pm–6:30pm	Plenary Session: Introduction and Warm-Up Welcome Overview of Program Discussion of Team Mentoring A Word from Our Sponsors (NSF)
7:00pm–9:00pm	Dinner
Day 1	
8:30am–9:30am	Breakfast
9:00am–9:30am	Logistics for Today
9:30am–11:00am	Team session #1 Each team session involves intensive discussion of one of the participant's work including its importance and contributions, improvements that could be made, likely publication strategies, and future research that might be attempted.
11:00am–12:00pm	Plenary Session Don't Perish: Research and Publishing This panel discussed tips on doing research, and also how the publishing process works. Also, covered topics like when to push back on an editor, and how to respond to a referee report.
12:00pm–1:00pm	Networking Lunch (seating by geographic location)
1:00pm–2:30pm	Team session #2

2:30pm–3:30pm	Plenary Session: Show Me the Money: Getting Grants This panel discusses strategies for getting grants. The NSF was heavily represented, but also other funding sources NIH, McArthur, private industry.
3:30pm–5:00pm	Team session #3
5:00pm–6:00pm:	Plenary Session: Exposing Yourself: Other Professional Activities This panel covered issues like giving talks at schools, giving talks (and discussing) at conferences, face-time and pseudo-social events, sending out work to colleagues, networking at conferences and in other forums. Also how to behave within your department, asking for a raise, interacting with the chair.
7:00pm–9:00pm	Dinner
	Day 2
8:30am–9:30am	Breakfast
9:00am–9:30am	Logistics for Today
9:30am–10:30am	Plenary Session: Complement or Substitute: Teaching This panel discussed tips for turning teaching from a substitute for research to a complement. Topics included discussing your research in the classroom, designing projects and other educational tasks for your students that help your research. Also, tricks for handling particularly difficult teaching loads (e.g., very large classes, mathematical and otherwise unpopular classes . . .)
10:30am–12:00n	Team session #4
12:00pm–1:00pm	Networking Lunch (seating by teaching interests)
1:00pm–2:00pm	Plenary Session: Clearing the Hurdle: Getting Tenure This panel presented information on how the tenure process works (this will be different at different schools, of course, but generally). Panelists also offered advice on aspects like choosing your research area (not too big, not too small), your reference group, etc.)
2:00pm–3:30pm	Team session #5
3:30pm–4:30pm	Plenary Session: Walking the Tightrope: Work-Life Balance This panel covered strategies for balancing one's work with one's life. This included dual career issues (trade-offs and sacrifices), children, and other balancing issues.
4:30pm–5:00pm	Summary and Follow-up, Adjournment

As suggested by the group structure and workshop design described above, workshops were developed to facilitate peer networking. Each participant was asked to submit a recent working paper, grant proposal, or other piece of work on which they would like comments. These were circulated in advance of the workshop, and each participant was asked to read and prepare comments on the work of her group members. For some participants this was the first time they had been asked to provide a formal set of comments on another researcher's work, and it provided them with great insight into the article review process. Additionally, since the regional workshops focused equally on teaching, those participants were asked to circulate a teaching exercise on which they would like to receive feedback. Participants were expected to review these documents for discussion at the workshop.

At the national workshops, much of the time was spent in small groups, discussing each other's research. Each participant received an hour of feedback on her paper from her team and the senior mentors. Everyone involved reported that this was an extremely important part of the workshops; not only did each participant get focused and intense feedback on her own work, it set a norm for future interaction among the group members of reading each other's work and providing comments. It kept the workshop grounded in research, rather than turning it into two days of receiving advice. Finally, discussions about particular papers provided a jumping-off point for the senior mentors to talk about streams of research in the specific field, journals that might be appropriate homes for this type of work, and other projects that were ongoing that may have been previously unknown to participants. These sessions served to develop relationships between the junior faculty and the senior faculty, and to develop the peer network structure that would continue after the workshop had ended.

Shorter versions of this research-based discussion were conducted at the regional workshops. Other small-group sessions in those workshops were designed to facilitate collective development of practical strategies for success in the areas of teaching and networking. For example, participants at the regional workshop were asked to complete a networking exercise explicitly designed to broaden their thinking of where networks could be developed. Each small group was required to come up with at least three networking strategies at conferences, in their specific field, and at their home institution.

Like the peers in the group, the mentors read and provided comments on the participant's work. They also provided a broader perspective than the peers for each participant, drawing on their years of experience on editorial boards, promotion and tenure committees, and in the field. Both junior and senior mentors played this role.

In addition to the small-group meetings at both workshops, we held plenary sessions consisting of panels of the senior mentors. Topics included research and publishing, getting grants, professional exposure, teaching, the tenure process, and work-life balance. For each topic, three to four senior mentors served on the panel, and offered five to ten minutes of their thoughts. The rest of the time was spent on questions from the floor.

These plenary sessions were extremely popular and helpful. They included tips and explicit advice in areas where academics usually receive little formal training (e.g., how to respond to a referee report, how to network at a conference, how to manage a particularly difficult teaching load). These sessions primarily served as vehicles for information transmission. In retrospect, we wished we had videotaped these sessions, as the collective knowledge that was represented would have been useful for those who could not attend.

At the end of the workshop, groups were instructed to make a plan for keeping their group together. Most chose to send monthly email updates, serving as commitment devices to stay on track, and offering suggestions on each other's work. Some groups made plans to submit paper sessions to upcoming conferences, representing their own closely related work. One ambitious group made plans to develop a seminar series across their institutions. Many collaborations were begun.

We also encouraged participants to rely on their group and their mentors in the future when they needed advice, including reading drafts of papers, responding to referees, discussing teaching quandaries, and asking for recommendation letters. We suggested that they invite each other to their home institutions to give talks, and that they meet periodically at conferences for lunch or other events. Subsequent reports from workshop participants suggest that many (but not all) of these groups stayed together in the way we had hoped.

In addition to within-group bonding achieved through the consistent group membership throughout the workshop, we also tried to encourage intergroup discussion and fertilization. Participants shared hotel rooms with someone from another research group. Dinners and lunches were explicitly across-group (organized by teaching, by geographical region, or by stage-of-life [single, married/attached no kids, with kids], for example). These efforts were somewhat less effective; at the end of the workshop individuals knew their group members quite well and other participants only marginally.

We had a number of implementation challenges as well. Finding mentors who were willing to dedicate their time and energy, both during the workshop and afterward, was no easy task. It was particularly difficult to find senior women in specific areas of economics (for example, economic theory and econometrics). And there were a number of excellent male colleagues who wanted to be involved. In the end we decided that the role model function of the mentor was important enough that we would insist on having female mentors, but in the future we may move toward a model of one male and one female mentor.

The biggest challenge for us was the work-life balance session. We originally focused this session on child care, but that topic was relevant only to a subset of the participants. Much of the advice, as well, was individual-specific. In order to provide participants with a better understanding of the diversity among women in their choice of relationships and their associated challenges and solutions, the most recent regional workshop session included three panelists with very different family structures. One panelist had three children while untenured, another waited until after tenure to have her only child, and the final was in a nonheterosexual relationship and had chosen not to have children.

Evaluation Efforts

At the end of the workshops we offered an exit survey. On a scale of 1–7 where 1 is "not at all helpful" and 7 is "extremely helpful" the average rating of the workshops were 6.69 (ninety evaluations received out of ninety-eight participants). The lowest-rated session (and the one that we are still refining) was the one on work-life balance.

In addition, after the workshop a number of participants emailed us about their experiences. Here are some excerpts from those emails:

I learned a lot from the workshop and I wish I would have attended 2 years ago.

I had a really fantastic experience at the workshop. So much information and networking packed into the 2 days!

I have a list of things I should be saving for my tenure file, ways to get my name out there, strategies that will help me get published, get tenure, and stay sane in the process.

The workshop provides junior faculty with an instant support group of people with similar research and teaching interests. Members of my research group send e-mails on a monthly basis, and we are planning sessions for two future meetings.

In addition to these informal evaluations, we are undertaking a more formal evaluation of the program based on the random assignment selection mechanism. As noted above, when individuals applied to attend the national workshop, they were first screened for qualifications and appropriateness, then were randomly selected to attend. In followup research we are collecting vitas from those who were assigned in and out, and will be comparing them on a number of dimensions, such as research productivity, conference attendance, and academic talks. We intend to compare these and other outcome measures to evaluate the effectiveness of the intervention.

Beyond Economics: What's Generalizable?

We believe that interventions containing the four aspects of mentoring *(role models, information transmission, informal relationships,* and *peer networks)* will be effective in any discipline or across disciplines within an institution. Here are some lessons we learned from our workshops, and the action–items that they suggest.

Clear Communication Is Often Lacking

One of the most surprising insights we gathered from our workshops is that participants don't know what is required to get tenure at their institution! Of course, all participants know the list of top and prestigious journals in their field, but no one could tell us how many publications in those journals would be necessary for tenure, or what the trade-offs were between fewer publications in top journals and more publications in lower-tier journals. Similarly, very few participants knew how important (or unimportant) it was to get grants, good teaching ratings, or do good service. In brief, the criteria for promotion were extremely unclear. One of the big payoffs for the workshop participants was getting additional insight (admittedly, from senior women not at their institutions) about what "counted" toward tenure, and roughly what the standards were.

Note that this is likely a problem with both men and women. However, men tend to have more same-sex colleagues, and may therefore be more comfortable asking more questions and gathering more information than women. Thus they are more likely to become informed on their own. Any interventions that increase communication on these dimensions will help all junior faculty, but especially women. We especially encouraged the participants to begin discussions with the senior faculty in their departments about these issues, and to strive for clarity in the requirements.

Increase Advice-Giving and Professional Conversations

For many of our participants, the discussions about journal editors, responding to referee reports, and other details of the publication process were their first. This information is critical for professional success, yet is not taught in Ph.D. programs or in typical hallway discussions. Participants seemed to have an image of research as a solitary activity; something done behind closed doors. We encouraged them to approach their newfound mentors, peer networks, and existing colleagues at multiple stages in the research process, asking them to read papers early, recommending journals, and asking for advice on responding to referee reports.

Similarly, most academics in economics do not have formal training in pedagogical practices. While many of the issues that arise are institution specific, women face a more general challenge. Students tend to typecast faculty, and it is commonly felt that men "get away with" acting more authoritarian than women. Many younger students still look to their female faculty for nurturing, both in and out of the classroom. Finding a balance between time spent on teaching and research might be especially challenging under these circumstances, and discussions regarding teaching styles and time savers are invaluable to junior faculty.

Again, advice from senior colleagues is likely to be useful for both men and women, but women seem less likely to get it. Creating relationships between individuals (whether mentors or peers) and expectations that these types of topics will be discussed can help academics in any field, and across a campus as well.

Focus on Professional Activities

Our participants severely undervalued professional activities other than publishing. Senior mentors were quite explicit about the need for sending one's papers to colleagues (before, during, and after the publication process), attending conferences, and becoming "known" within your field, as well as other networking activities (termed by one of our mentors "quasi-social events"). These activities don't have direct payoffs, but do have large indirect payoffs; for example, if an audience member from a conference where a paper is presented later evaluates it for publication, it already will be familiar.

Again, this bias seems to stem from the vision of the researcher as solitary; the thinking is that the tenure decision is based solely on looking at one's vita. But in fact, schools want to know whether a tenure candidate is well known in her field, and is poised to become a leader. The quality of the tenure letters received, the senior mentors argued, is also a function of these professional activities. While this effect may differ between disciplines, activity at conferences seems undervalued by junior faculty, and information about its importance is worth transmitting.

Similar Programs in Other Disciplines

Mentoring programs exist in a number of disciplines but vary in both their focus and the process by which they achieve their goals. For example, the Committee on the Advancement of Women in Chemistry (COACh) originated as a leadership program for tenured women and has recently expanded the target audience to the untenured. (For more information see http://coach.uoregon.edu/coachfiles/workshops .html.) The Committee on the Status of Women in Computing Research has also focused its efforts on the more senior levels. Their Cohort of Associate Professors Project (CAPP) brings together associate professors with more senior mentors for two-day development workshops on advancing careers. (For more information see http://www .cra.org/Activities/craw/capp/index.php.)

Other disciplines have made their mentoring activities entirely electronic. For example, the Women in Linguistics Mentoring Alliance (WILMA) has a self-sustaining web site where junior faculty log in and search for registered senior faculty who share their interests (the web site can be viewed at http://ling.wisc.edu/wilma/). Similarly, MentorNet electronically pairs engineering undergraduates, graduates, and junior faculty with professional engineers or senior faculty to help them advance their careers (http://www.mentornet.net/).

Our CeMENT mentoring workshop for women in economics differs from these efforts in a number of ways. The national workshops devote a significant amount of resources in discussing and providing substantive feedback on the research projects of participating women. The team sessions are designed to provide feedback on a specific project with the intent of enhancing that work in a way that increases the probability of being published in a top-tier journal. Further, both national and regional workshops make use of teams in anticipation that conversations will continue beyond this one-time intervention. Finally, CeMENT workshops are more diverse in the topics covered, providing women the opportunity to address a number of issues (that often overlap) in a single event.

Challenges Implementing across Disciplines, within an Institution

While discipline-specific programs have benefits such as providing a venue for vetting current research projects, communicating journal-specific information, and offering teaching enhancements related to content, they run the risk of having participants return to the isolation of home institutions without a local support system.

Thus we believe that institution-specific programs complement the efforts of disciplinary programs and can provide a basis for long-term sustainability. Indeed, one of the lessons that emerges from our workshops is the importance of finding mentors and supporters at the home institution. We encourage our participants to rely not just on the relationships they form within economics, but to reach out to others at their institutions who might help them or whom they might help.

However, institutions face special challenges in implementing mentoring or networking relationships across disciplines, as a result of the often-surprising heterogeneity in publication types and expectations in various academic areas. Furthermore, there may not be enough senior female faculty to serve as mentors for the existing junior faculty within a given institution. Similarly, there may be a lack of critical mass within a given discipline to form peer networks. Administrators will need to think creatively to identify solutions to these challenges. Here we offer our ideas, but stress that they are far from exhaustive.

Get Outside Help

Bringing leading researchers and pedagogical specialists to campus to give talks is common practice. It is a small matter to target one or two female seminar speakers each year in each department, and they can serve as important role models for the institution's female junior faculty. The university can to ask these visitors to stay an extra day, to meet with the female junior faculty to discuss research, give information, and create relationships.

Create Peer Networks, within and across Disciplines

Institute programs to encourage junior faculty within departments to get together and discuss their research and teaching (this might be as simple as providing lunch for an internal seminar series, or as complicated as planning a two-day retreat). Encourage them to call each other with problems, challenges, or celebrations of positive events. Across disciplines the focus naturally moves away from the research and toward challenges that junior faculty all share. This can include teaching, professional and institution-specific service, dealing with the institution's administration, and understanding the tenure process.

Create Mentoring Relationships

It will likely be infeasible for each female junior faculty to have a female senior faculty within her department (much less her specialty) to serve as a mentor. However, this role can be taken on by more than one person. For example, each junior faculty could be assigned a senior person within her research area for research advice, a distinguished teacher at the institution for pedagogical advice, a female senior person in a different department for gender-specific perspectives, and possibly a senior person outside the university for an outside perspective. Thinking creatively about what relationships exist and which are needed for each individual will help create the relationships needed for academic success.

Conclusion

Our objective was to increase the tenure rate for women in economics to the rate we observe for men, and to bring economics to parity with other sciences. We believe this improvement will benefit not just the participants in our workshops, but the field as a whole. It will increase the perceived and actual promotion likelihoods for women, attracting more and better-qualified women to economics as a career. In addition, it will increase the number of tenured female faculty in economics, who can then serve as role models in their own right for future generations of junior women.

By bringing together junior and senior women, participants gained from the *role model* function of mentoring, often absent because of low numbers of senior women in any given department. Second, junior and senior researchers established *informal relationships* that continued past the workshop. Third, by explicitly running sessions and offering advice on publishing, teaching, career advancement, and work-life balance, participants received the *information transmission* benefits of mentoring. Finally, by using the teams, we built *peer networks* that we believe are crucial to the success of academics in any field. However, we believe there is yet a fifth benefit from these workshops, which has to do with *role activation* and intergenerational altruism.

When one speaks to senior faculty who are dedicated to helping others succeed, they often mention that their dedication stems from the mentoring they themselves received. We believe that one benefit from these workshops will be to activate in the participants their identity not just as academics, but as female academics. We believe we can then use this activation to help other women in the field. We have informally observed this activation from the previous workshops—all of the participants contacted have enthusiastically volunteered to serve as midcareer mentors in this upcoming round of workshops. In addition, many participants in the first workshop have "taken the lessons home" by helping junior women in economics and in other fields at their institution. We believe that institutions need to remain dedicated to activities like these over time, which will encourage participants in early workshops to reciprocate the benefits they received.

In this chapter, we have described the workshops we ran, the lessons we learned, and how those lessons can inform other professional organizations or institutions in their own efforts. We have included some specific suggestions for how to adjust the content and format of workshops for different fields, and for interdisciplinary audiences. We hope this material can serve as a starting place for similar interventions by other organizations, to help achieve our shared goal of appropriate female representation in academia.

REFERENCES

Blank, R. 1993. Report of the committee on the status of women in the economics profession. *American Economic Review* 83:492–95.

Blau, F. 2002. Report of the committee on the status of women in the economics profession. *American Economic Review* 92:516–20.

Blau, F., M. Ferber, and A. Winkler. 2002. *The economics of women, men, and work.* 4th ed. Upper Saddle River, NJ: Prentice-Hall.

Committee on the Status of Women in the Economics Profession. 2002. Report of the Committee on the Status of Women in the Economics Profession. *AEA Papers and Proceedings* 92 (2): 516–20. http://www.cswep.org/annual_reports/2001_CSWEP_Annual_Report.pdf.

Ginther, D., and S. Kahn. 2004. Women in economics: Moving up or falling off the academic career ladder? *Journal of Economic Perspectives* 18:193–214.

Kahn, S. 2002. The status of women in economics during the nineties: One step forward, two steps back. Paper presented at the American Economic Association Meetings, Atlanta, January.

McDowell, J., and M. Melvin. 1983. The determinants of co-authorship: An analysis of the economics literature. *Review of Economics and Statistics* 65:155–60.

McDowell, J., L. Singell, and M. Slater. 2002. Two to tango? Gender differences in the joint decision to publish and coauthor. Paper presented at the American Economic Association Meetings, Atlanta, January.

McDowell, J., and J. Smith. 1992. The effect of gender-sorting on propensity to coauthor: Implication for academic promotion. *Economic Inquiry,* 30:68–82.

Nelson, Donna J. 2005. A national analysis of diversity in science and engineering faculties at research universities. http://cheminfo.chem.ou.edu/~djn/diversity/briefings/Diversity%20Report%20Final.pdf.

Ragins, B. 1999. Gender and mentoring relationships: A review and research agenda for the next decade. In *Handbook of gender and work,* ed. G. Powell, 347–70. Thousand Oaks, CA: Sage.

Zachary, L. 2000. *The mentor's guide: Facilitating effective learning relationships.* San Francisco: Jossey-Bass.

Zimbler, Linda J. 2001. Background characteristics, work activities, and compensation of faculty and instructional staff in postsecondary institutions: Fall 1998, NCES 2001–152. U.S. Department of Education, National Center for Education Statistics. http://nces.ed.gov/pubs2001/2001152.pdf.

ELEVEN

Equity in Tenure and Promotion

AN INTEGRATED INSTITUTIONAL APPROACH

Mary Frank Fox, Carol Colatrella, David McDowell,
& Mary Lynn Realff

THE GEORGIA INSTITUTE of Technology (Georgia Tech) NSF ADVANCE initiative takes an "integrated institutional approach" to factors that support the advancement of faculty and provide a model of best practices in academic science and engineering. With this approach, the emphasis is upon organizational factors and features that have been shown to shape positive outcomes of participation and performance for women in science and engineering. These organizational factors include leadership and organizational climate that signal the importance of equity (Fox 1998, 2000; Valian 1999), clarity and equity of guidelines for evaluation and rewards (Evetts 1996; Fox 1991, 2000; Long and Fox 1995), collegial and collaborative networks of access and opportunity (Fox 1991; Reskin 1998), and supportive family policies (Raabe 1997; Rosser and Zieseniss 2000).

Accordingly, the Georgia Tech ADVANCE initiative approaches participation, performance, and advancement as organizational issues, subject to institutional transformation. Because processes of evaluation are critical to positive outcomes for faculty in science and engineering, they are an indispensable part of institutional transformation. Enhancing fair processes of evaluation entails understanding theory and research on social conditions associated with equity in evaluation; and implementing practices and policies that support those conditions. In this chapter, we describe the underlying theory and research, and the practices and policies developed and implemented to support equitable evaluation as a key part of Georgia Tech's ADVANCE initiative.

Bias and Equity in Evaluation

Theory and Research

The normative standards applied to scholarly and scientific work can set the stage for bias in evaluation: "In this work, standards are both 'absolute' and 'subjective.' Performance is measured against a standard of absolute excellence, which in turn, is a subjective assessment" (Fox 1991, 191). When the criteria—and processes—of evaluation are subjective, loosely defined, and a matter of "judgment," bias in assessment is more likely to result (see Blalock 1991; Evetts 1996; McIlwee and Robinson 1992; Reskin 2003; Tierney 1995). Three sets of conditions, in contrast, support equitable evaluation (Long and Fox 1995, 62–64).

1. More complete information on candidates' records and qualifications: When more, compared to less, information is available on candidates' records and qualifications, considerations of race, gender, and national origin are less likely to be brought to bear in the evaluation. "Non-performance-based characteristics," such as gender and race, are more likely to be activated as bases for evaluation when there are few relevant and known criteria on which to judge individual performance.

2. Clarity of standards: When criteria for evaluation are ambiguous, outcomes based upon race, gender, national origin, and other personal and social characteristics are more likely to occur. Studies indicate that the more loosely defined and subjective the criteria, the more likely it is that persons with "majority group characteristics" will be perceived as the superior candidates, and that bias will operate. When criteria for evaluation are more standardized, bias in evaluation is reduced.

3. Open processes: Open processes in hiring, promotion, and allocation of rewards increase more equitable decision-making, as does the exercise of due-process procedures. Secret and nonsystematic processes tend to activate "particularistic considerations"—that is, characteristics such as race, gender, national origin, levels of physical ability, and other personal and social attributes—in evaluation.

Practice

The Georgia Tech Promotion and Tenure ADVANCE Committee (PTAC) introduced four sets of initiatives to support equity in evaluation for promotion and tenure (PTAC 2003): (1) a comprehensive canvass of current tenure and promotion processes; (2) the PTAC survey of faculty; (3) specification of "best practices"; and (4) the Awareness of Decisions in Evaluating Promotion and Tenure (ADEPT) instrument. These four initiatives reflect theory and research on equity in evaluation, and represent an integrated set of institutional approaches toward equity in tenure and promotion.

COMPREHENSIVE CANVASS OF EVALUATION PROCESS,
ACROSS UNITS

PTAC canvassed the current tenure and promotion practices across all units, with an aim to document the range of procedural practices among academic units at Georgia Tech, and to compare these practices. This comprehensive gathering of information included methods of committee assignment by unit; constitution of committees; structure of processes at the unit level; expected information from candidate, including format; faculty performance expectations; quality of publication and presentation venues; provisions for feedback to candidate; and mentoring.

This canvassing pointed to variation among units of both written and unwritten guidelines as to how evaluation is conducted. Such variation reflects the history of Georgia Tech as an entrepreneurial institution with a decentralized hierarchy in which chairs exercise a great deal of control over their units. This system has encouraged units to develop missions, procedures, and standards reflective of their areas or disciplines while also adhering to the mission and goals of Georgia Tech as envisioned by central administration. A consistent theme of the PTAC report is that written guidelines developed to supplement the university system regulations and Institute handbook should exist, especially at the unit level, to set expectations for the performance of faculty and for the committees undertaking the reviews.

PTAC SURVEY

The PTAC committee acknowledged that understanding (and much more, enhancing) institutional culture regarding promotion and tenure evaluations would require the input of individual faculty as well as that of unit chairs and evaluation committee chairs contributing to the canvass of written and unwritten evaluation guidelines and practices. Thus, the PTAC web-based survey of academic faculty was undertaken,[1] focusing on issues pertaining to evaluation, including resource allocation and success; mentoring and networking; perceptions of evaluative methods and procedures; interdisciplinary collaboration; entrepreneurship; and institutional culture. The survey resulted in 325 responses, representing 37% of the academic faculty of the institution. A detailed presentation of the survey and findings appears at http://www.advance.gatech.edu/ptac/.

Primary findings include these:

1. Resource allocation: Faculty reported that "support for graduate students" and "quality laboratory equipment" were the two most important resources for success of faculty.

2. Mentoring and networking: Faculty perceived mentoring as important toward success in promotion and tenure, with women faculty rating mentoring as more important than did men. Respondents also considered networking with peers, both inside and outside of Georgia Tech, to be important for successful promotion and tenure.

3. Perceptions about methods and procedures for evaluating research, teaching, and service: The faculty believed that research records were subjected to a higher level of scrutiny than teaching records, and that service records had the lowest level of scrutiny. When faculty were asked what they consider to be the "most effective methods of measuring research productivity," number and quality of journal articles, quality of conference papers, and number of Ph.D. students were reported as "most effective." The mean level of response of understanding the tenure and promotion process was 3.28 (on a four-point scale, where 4 = understand very well), with no significant differences by gender, college, or administrative status.

4. Interdisciplinary collaborations: Faculty believed that interdisciplinary collaboration was valued "moderately"—but the consistency in evaluation was low for this category.

5. Entrepreneurship: The value of entrepreneurial activities in home units was reported to be "low" in home units. However, faculty reported that except for promotion and tenure reviews, all other support or evaluative systems at Georgia Tech "somewhat encourage" entrepreneurial activities.

6. Environment and culture of the institution: Most reporting faculty said that they had not been harassed or experienced comments regarding their personal appearance or attire. When asked to compare the culture of Georgia Tech to peer institutions with regard to being "progressive," faculty rated it as about the same in all aspects *except* "time to reflect and write" and "staff and infrastructural support," which were perceived to be slightly behind/about the same as peer institutions.

The PTAC survey serves as a benchmark of faculty perceptions, a baseline for continuing assessment, and a reference point for the further development of materials to assist evaluation candidates and committee members. The PTAC survey data confirm anecdotal impressions of Georgia Tech as a competitive environment focusing on research, and the survey results articulate faculty needs for material resources and time to enhance their research programs. Faculty respondents, particularly women, noted their interest in receiving mentoring. The faculty responding to the PTAC survey report that their understanding of the criteria for evaluation is "moderately clear." At the same time, however, interviews with women faculty, conducted by the ADVANCE research program, point to the strong perception that the *application* of criteria for evaluation "varies with the candidate"—that is, that criteria for evaluation are applied nonuniformly (Fox and Colatrella 2006). The PTAC committee has responded to articulated faculty needs by developing the best-practices document and the ADEPT instrument, both described below.

RECOMMENDED BEST PRACTICES

The PTAC initiative developed *Recommended Best Practices in Reappointment, Promotion, and Tenure Processes* as part of its canvass of procedures across units, a study of the existing hierarchical structure for promotion and tenure evaluations at Georgia Tech, its survey of faculty, and review of theory and research on equity in evaluation. The best practices in evaluation, resulting from this integrated, institutional approach, include

- open and transparent processes, with clarity of expectations expressed in written documentation by colleges and home units, and with consistency with expectations expressed by the Institute;
- clarity of rationale, expectations, and operating guidelines for evalu-

ation committees, and consistency in application of criteria, across all levels of the review (home unit, college, Institute);

- clarity and completeness of candidates' contributions, as expressed in their documentation, including clear explanations of their roles in coauthorship of articles, collaborative activities, entrepreneurial activities, development of new and innovative educational programs or research initiatives, and involvement in professional societies and leadership opportunities—as well as identification of their "top five intellectual products";

- mentoring and faculty development, with emphasis upon communication and development of candidates' "personal plans" in research, teaching, and service as a "road map" to assist in prioritizing activities and setting timelines and benchmarks.

The web site http://wwww.advance.gatech.edu/ptac/ provides a detailed presentation of the best-practices document.

ADEPT INSTRUMENT

The fourth PTAC initiative is a web-based instrument, Awareness of Decisions in Evaluating Promotion and Tenure (ADEPT). This computer instrument includes (on the web site http://www.adept .gatech.edu) case studies of fictional candidates, downloadable narrative games linking educational practices and research, references to scholarly and news accounts of equity and bias in evaluation, and links to Georgia Tech resources on promotion and tenure.[2] The instrument is intended for use by candidates for promotion and tenure and members of unit-level committees who are evaluating promotion and tenure cases in U.S. universities and colleges. The primary goals of the ADEPT site and instrument are to assist users in identifying forms of bias in evaluation processes and to promote fair and objective evaluations.

The case studies are fictionalized career accounts of nine individuals in different disciplines, illustrating issues deemed significant by PTAC. The case studies reference the written and unwritten guidelines provided to candidates and the best practices of units and colleges. The issues in the case studies include sexism, racism, ethnic bias, and bias exhibited against those with disabilities, as well as biases associated with leaves of absence, interdisciplinary scholarship, publication venues, spousal hires, international reputations, joint appointments, "hard" ver-

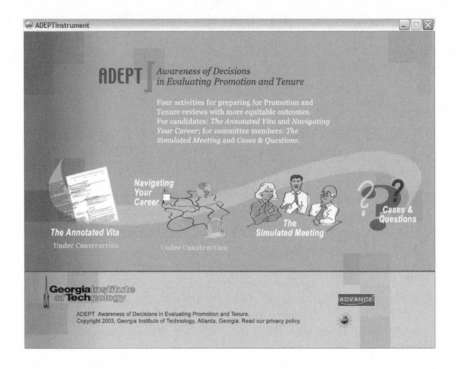

sus "soft" research, and more. PTAC members researched these issues and contributed relevant summaries of them to the PTAC report online at http://www.advance.gatech.edu/ptac/.

In 2002, the ADEPT design team began building an evaluation tool based on narrative components for Georgia Tech that would inform both faculty candidates and new committee members, demonstrate how eval-uation processes are conducted, and persuade committee members that individual and group judgments may be skewed by social bias and unstated professional preferences.[3] Although proceeding from some specific Georgia Tech concerns—including shifting standards accorded interdisciplinary scholarship, assessment of fluctuating productivity, response to health or parental leaves, discriminatory attitudes toward women and minorities, and the meaning of "early tenure"—the ADEPT tool also reflects common values and issues of research universities.

In proposing that attention be directed to the promotion and tenure process, the ADVANCE team acknowledged long-standing concerns at Georgia Tech that female, compared to male, faculty in science, mathe-matics, computing, and engineering have been subjected to greater scrutiny in review processes, and that largely male evaluation committees may be reluctant to tenure and promote women. Two Georgia Tech reports issued in 1998 pointed out that women faculty were more likely than men to leave the university before the end of the probationary period, and women anecdotally reported being discouraged from coming up for evaluation or not being appropriately advised about evaluation procedures and policies.[4]

Drafts of the fictional career accounts, dialogue, and representative (but fake) vitae incorporated revisions suggested by PTAC members.[5] This collaborative process helped to develop the career accounts and the activities as realistic and relevant to promotion and tenure issues.

Carol Colatrella, the primary writer and coordinator of ADEPT, worked to infuse research from social studies of science into the case studies and narrative games in the instrument, including references from literature in the annotated bibliography. For example, research has shown that establishing clear guidelines and broadly communicating these standards in a formalized manner helps women and minorities to fare better in the evaluation process. Clear, consistent communication to candidates and among committee members reduces bias in evaluation outcomes. Such guidelines and standards are embedded in ADEPT activities, which are further connected to an annotated bibliography of research focusing on a broad range of forms of bias related to gender, minority status, choice of publication venues, preference for interdisciplinary research, assignment of service activities, allocation of resources, mentoring, and disability.

The current beta 1.5 version of the ADEPT instrument includes nine case studies providing narrative career accounts of fictional candidates, three fictional promotion-and-tenure meetings inviting user participation and reflection ("Simulated Meeting" and "Follow Up Analysis"), and one version of a career strategy game ("Navigating Your Career"), plus a library of scholarly and Georgia Tech resources.

The fictional cases in the instrument allow candidates, committee members, administrators, and staff to discuss personnel issues without fear or anxiety that the actual subject of the dossier is in the room or might learn of such discussions. Because details of the cases are drawn from research at Georgia Tech, scholarly literature on bias in evaluation, news accounts, legal cases, and other sources, the resulting career accounts are realistic and provoke genuine disagreement. Thus, in 2003, the fictional cases formed the basis for breakout groups at the Georgia Tech ADVANCE conference, bringing together largely female faculty members and senior administrators (almost all male).

The same cases were considered in different breakout groups that mixed senior and junior faculty, resulting in very different decisions cast by different groups acting on the same information. The case studies focus on issues that often perplex evaluation committees, such as how to treat a candidate's probationary leave, fluctuating productivity, or interdisciplinary activity. Such issues are viewed differently by faculty according to their own backgrounds, experiences, and perceptions of equity. For example, the case study describing Patty Shen in Biomedical Engineering provoked divergent reactions among those at the conference who recognized the candidate's achievements as varying over time. Although productive in her first three years and the recipient of an Early Career Award, Shen gave birth to a baby in her fourth year and experienced consequent medical problems in that year and the following one that dramatically affected her publication record. Discussants differed in assessing the likelihood of Shen's future productivity. A number of people expressed the idea that early promise is not always sustained, and that Shen should have been able to publish more as she had taken a leave for one year and been released from teaching during one term, benefits that gave her "more time" than other candidates had. Other discussants, who supported awarding Shen promotion and tenure, asserted that long-term medical difficulties in her fourth and fifth year explain her lack of productivity and do not make her "a bad bet," as demonstrated by her bouncing back to publish more articles in the year before coming up for review. The conference plenary session reviewing the breakout discussions reported different decisions by different groups on Shen's case, some for and some against awarding her promotion and tenure.[6]

As social studies of evaluation have shown, assessments about the participation and performance of women—and men—in science and engineering departments are dependent on social factors, including features of the departments and institutions in which faculty are located (Fox 2000; Fox and Colatrella 2006), as well as social psychological perceptions about women and men (Valian 1999). Literature on evaluation and anecdotal reports from men and women faculty members indicate that few promotion and tenure dossiers can be regarded as absolutely stellar or failed. Instead, most dossiers are ambiguous on the surface. Therefore, if there is a will to do so, candidates' records can be cast as positive cases or negative (Fox 1991). The same details might even be pointed to as "facts," "evidence," to describe the dossier as either positive or negative, depending on that will. ADEPT cases and interactive exercises challenge

individuals to articulate their standards, with an aim that continuing conversation, consideration, and clarification will lead to more equitable practices.

The "Simulated Meeting" activity asks learners to participate in fictionalized promotion-and-tenure committee meetings, so that committee members assess and address issues that also affect real-world candidates under evaluation. Activities are built in Macromedia Director with Flash animations. "Simulated Meeting" consists of a hypermedia dialogue among three fictional characters and the user. After the user reads through the career account, he or she can read details about committee members, along with a schematic vita of the candidate, before moving through the dialogue of the fictional committee meeting. Statements regarding the candidate under consideration from the committee meeting are archived in the left column, while a cartoon graphic representation of the committee members occupies the right column with each new statement from a committee member indicated in a balloon.

Users are prompted to maintain an active engagement with the emerging dialogue. Each user of a simulated meeting has three opportunities to select statements that respond to arguments articulated by the three fictional committee members. The user's choices affect the meeting's out-

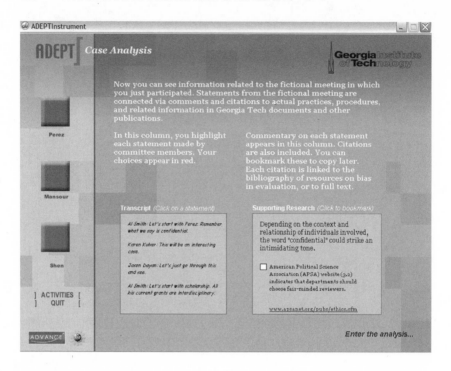

come in that other committee members react to the user's statements and proceed differently depending on which choice the user makes.

Each simulated meeting also has a follow-up analysis activity, which links the fictional issues to the bibliography of scholarship on bias in evaluation. Comments and citations for many statements uttered in the committee meeting appear in the right column, as the user highlights portions of the transcript he or she built. Comments and citations are linked to web addresses or the annotated references in the bibliography of bias in evaluation. So far, three different meetings are built for three cases (Perez, Mansour, and Shen); each meeting has a follow-up analysis.

The prototype of "Navigating Your Career," an activity designed for candidates and mentors, is built for the "Pam Lee case." After reading the associated career account, the user clicks on the yellow dot appearing somewhere on a map linking university, home, and the world. Each yellow dot represents a random decision point for the user, who responds to the decision by choosing among the options provided. After responding, the user reads an explanatory note considering the chosen course of action in relation to other options. User choices are scored in columns indicating research, teaching, service, health, and allies, with

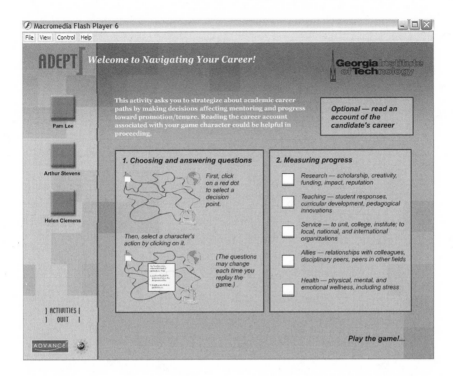

the last response generating the final score. The game can be played numerous times with different decision points generated and different scores possible. It is intended to stimulate discussion between junior and senior faculty, to supplement mentoring juniors might receive from senior colleagues, and to encourage faculty to think strategically about managing their careers.

"The Annotated Vita" activity is under construction. It will provide users with the opportunity to test different ways of formulating information on a vita for submission for critical review or promotion and tenure. To demonstrate the benefits of different formats for different disciplines, the vita activity will show how some candidates described in the case studies might construct their vitae toward positive or negative outcomes.

Implementation of ADEPT

The ADEPT design team rolled out the beta 1.0 version of ADEPT in spring 2003; the improved version 1.5 became available in November 2004 from the web site http://www.adept.gatech.edu. A PC user can receive the current activities in two separate downloads from that site; a Flash version to be posted at the ADEPT site is under construction. Users are asked to complete a short evaluation at the same site; the evaluation is designed to capture the usefulness of the instrument and to report any problems. Georgia Tech and other university faculty at two ADVANCE conferences (2004 and 2005) have previewed the instrument. Colatrella has also articulated the rationale of the promotion and tenure research and demonstrated the instrument for faculty at conferences and at particular campuses requesting previews.[7]

Georgia Tech's administration has directed colleges and home units to consult the Promotion and Tenure ADVANCE Committee report, particularly the section on best practices, and the ADEPT instrument, in preparing candidates and committees for reviews. The 2004–5 college review committees at the university reported successfully implementing consideration of the fictional cases and activities as preliminary to discussing the cases of real candidates. As Antonia Palmer (2005) notes, plausible narrative accounts can assist in management training: "Stories combined with learning activities that are situated . . . allow the learner to try out 'real' tasks, deal with 'real people,' and 'real situations.'"

Rather than training faculty to pursue a definitive course of action regarding any of the candidates, ADEPT encourages faculty reviewers to carefully consider cases to avoid exhibiting or replicating biases. ADEPT's construction allows different units to maintain disciplinary preferences in reviewing scholarly productivity, while being advised to avoid social bias, including that based upon gender and race or ethnicity. Discussion of the cases and activities encourages individuals to understand how their views represent or deviate from those of their peers.[8] ADEPT thus serves as a means for establishing clear, consistent communication about evaluation among candidates and committee members, as recommended in the PTAC "best practices" and the underlying theory and research on evaluation.

Written evaluations of ADEPT at Georgia Tech praise the instrument for replicating evaluation issues and providing information that is especially useful for junior faculty and faculty new to evaluation processes. Informal evaluations of the instrument from scholarly groups and other universities indicate "hunger" for information about processes rarely provided by one's home unit, the very unit in which the promotion and tenure case is evaluated at the beginning of the review process. Georgia Tech faculty members have also begun incorporating discussion of ADEPT cases and games into their mentoring discussions with faculty so that junior faculty and those new to the Institute might better understand expectations and procedures.

Although the instrument's content and design have been universally praised, many users have experienced significant technical difficulties with using Flash within a Macromedia Director platform. The interest in having the instrument's activities made more easily accessible has encouraged the ADEPT design team to build the games as Flash on the web. This conversion to a totally web-based instrument is in process.

ADEPT has been built for Georgia Tech, but disseminated through the coalition of nineteen universities funded through the second round of IT grants under the NSF ADVANCE initiative, and will therefore be adapted to a broader range of practices during its continuing development. When the final version becomes available, it will likely be maintained by the Center for the Study of Women, Science, and Technology, now sponsored by the Office of the Provost.

Policy

The theory and research and the practices of Georgia Tech's integrated institutional approach to awareness of both bias and equity in tenure and promotion have implications for institutional policies, including those:

- To establish guidelines and provide support for preparation of reappointment, promotion, and tenure packages ("dossiers").
- To provide guidance or mentoring for candidates in the review for the appointment, promotion, and tenure (RPT) process.
- To ensure that faculty members are properly respected and nurtured at Georgia Tech.
- To set an objective of written guidelines for unit-level RPT committees, especially, thus implementing "best practices."
- To equip faculty to identify and address forms of bias, even subtle forms, that may commonly enter into the process of evaluation.
- To reduce the extent to which bias enters the process of RPT.

Processes of appraisal of faculty reflect the "interests" of an institution, its objectives, expectations, and priorities (Wilson and Beaton 1993). Fair and useful policies of evaluation are likely to be successfully implemented when they are in concert with the goals, aspirations, and leadership of the institutions (Arreola 2000).

Summary

Georgia Tech's ADVANCE program takes an integrated institutional approach to factors that support the advancement of faculty and provide a model of best practices in academic science and engineering. With this approach, the emphasis is upon organizational features and factors that have been shown to shape positive outcomes for faculty, including women and underrepresented groups. Among these, a central organizational factor is "clarity and equity of guidelines for evaluation and rewards." Equitable evaluation involves understanding theory and research on social conditions that support fair processes of evaluation, and in turn, involves practices and policies that reflect those conditions.

In keeping with this emphasis, the Georgia Tech Promotion and Tenure ADVANCE Committee (PTAC) has introduced, developed, and implemented four key practices toward equity in evaluation: (1) a comprehensive, Institute-wide canvass of documents in tenure and promotion to determine the extent and nature of current guidelines; (2) a PTAC survey of faculty to determine areas of perceived needs; (3) specification of "best practices" that derive from theory and research, as well as the PTAC canvass and survey; and (4) the Awareness of Decisions in Evaluation of Promotion and Tenure (ADEPT) web-based instrument to assist users in identifying forms of bias in evaluation and in promoting fair and equitable processes of evaluation.

These practices have implications for institutional policies to establish guidelines for evaluation; provide guidance and mentoring for candidates of evaluative processes; ensure that faculty members are respected and nurtured; equip faculty with means to identify bias and reduce the extent to which bias enters the evaluative processes. The overall organizational framework for success of these policies rests on their interaction with—and their reflection of—Georgia Tech's aspirations and goals to be among the world's best technological universities, and to do so by supporting and recognizing outstanding accomplishments of its faculty.

NOTES

1. This web-based survey, led by PTAC chair Dr. David McDowell, was constructed and administered with guidance by Dr. Mary Frank Fox (co–principal

investigator of GT's ADVANCE program), and by Dr. Joseph Hoey (GT Office of Assessment).

2. Drafts of these materials are available at http://www.adept.gatech.edu and in an incorporated downloadable application.

3. Names of design team, ADVANCE team, and PTAC contributors are noted at http://www.adept.gatech.edu/team.htm.

4. The two prior reports are the "InGEAR Report on the Status of Women at Georgia Tech, 1993–1998," posted at http://www.academic.gatech.edu/study/, and the 1998 report of the Georgia Tech College of Engineering Task Force for Opportunities for Women in Engineering, "Enhancing the Environment for Success."

5. Drs. Carol Colatrella and David McDowell wrote the career accounts, dialogue, and fictional meetings in consultation with PTAC members.

6. Faculty, reporting on the discussions, agreed that speaking about the fictional Shen was a good thing to do as her case contained authentic elements resembling those that spark committee disagreement, and that reviewing a fictional persona permitted them to speak more candidly than they would about a real individual. It allowed them to engage in more fruitful debate and consideration of how to treat a candidate's leave, medical factors, variant productivity, and decision to wait until her second opportunity to come up for review.

7. Colatrella presented ADEPT cases and activities at the following conferences: the Society for Literature, Science, and the Arts (SLSA) in 2004; the Modern Language Association (MLA) in 2004, and the Society for the Study of Narrative Literature (SSNL) in 2005.

8. Because the web site and games based on the cases are highly interactive, they can be used profitably by individuals or groups. The simulated meetings allow a user to take the role of a committee member in a scenario that partly develops according to the user's interventions. Other games encourage candidates to consider career strategies, while senior faculty might consider how such strategies can be best conveyed to junior faculty they mentor.

REFERENCES

Awareness of Decisions in Evaluating Promotion and Tenure (ADEPT). http://www.adept.gatech.edu/.

Arreola, Raoul A. 2000. *Developing a comprehensive faculty evaluation system.* 2nd ed. Bolton, MA: Anker.

Blalock, Hubert. 1991. *Understanding social inequality: Modeling allocation processes.* Newbury Park, CA: Sage.

Evetts, Julia. 1996. *Gender and career in science and engineering.* London: Taylor and Francis.

Fox, Mary Frank. 1991. Gender, environmental milieu, and productivity in science. In *The outer circle: Women in the scientific community,* ed. Harriet Zuckerman, Jonathan Cole, and John Bruer, 188–204. New York: W. W. Norton.

———. 1998. Women in science and engineering: Theory, practice, and policy in programs. *Signs* 24:201–23.

————. 2000. Organizational environments and doctoral degrees awarded to women in science and engineering departments. *Women's Studies Quarterly* 28:47–61.

Fox, Mary Frank, and Carol Colatrella. 2006. Participation, performance, and advancement of women in academic science and engineering: What is at issue and why. *Journal of Technology Transfer* 31:377–86.

Long, J. Scott, and Mary Frank Fox. 1995. Scientific careers: Universalism and particularism. *Annual Review of Sociology* 21:45–71.

McIlwee, Judith, and J. Gregg Robinson. 1992. *Women in engineering: Gender, power, and workplace power.* Albany: State University of New York Press.

Palmer, Antonia. 2005. Case studies as interactive narrative: From traditional case studies to on-line case story interactive narrative environments. http://cloe.on .ca/documents/cloejul4antonia.ppt.

Raabe, Phyllis Hutton. 1997. Work-family policies for faculty. In *Academic couples: Problems and promises,* ed. Marianne A. Ferber and Jane W. Loeb, 208–25. Urbana: University of Illinois Press.

Promotion and Tenure ADVANCE Committee (PTAC). 2003. Report of the Georgia Tech Promotion and Tenure ADVANCE Committee (PTAC). Chaired by Dr. David L. McDowell. www.advance.gatech.edu/ptac

Reskin, Barbara. 1998. Sex differentiation and social organization of science. *Sociological Inquiry* 48:491–504.

————. 2003. Including mechanisms in our models of ascriptive inequality. *American Sociological Review* 68: 1–21.

Rosser, Sue, and Mirellie Zieseniss. 2000. Career issues and laboratory climates. *Journal of Women and Minorities in Science and Engineering* 6: 1–20.

Tierney, Margaret. 1995. Negotiating a software career. In *The gender-technology relation: Contemporary theory and research,* ed. Keith Grant and Rosalind Gill, 192–209. London: Taylor and Francis.

Valian, Virginia. 1999. *Why so slow? The advancement of women.* Cambridge: MIT Press.

Wilson, Fiona, and Beaton, Donald. 1993. The theory and practice of appraisal: Progress review in a Scottish university. *Higher Education Quarterly* 47:163–89.

TWELVE

Executive Coaching

AN EFFECTIVE STRATEGY FOR FACULTY DEVELOPMENT

Diana Bilimoria, Margaret M. Hopkins, Deborah A. O'Neil,
& Susan R. Perry

HIGHER EDUCATION researchers and university administrators alike are increasingly concerned about the persistent dearth of women faculty in higher education, the overall glacial advancement of women, and the existence of a glass ceiling in academic science and engineering (S&E) fields. The sources of these problems may be traced to individual psychological processes (gender schemas) and systematic institutional barriers, resulting in perceptions of a chilly climate for women scientists and engineers in academia (Sandler and Hall 1986), the experience of subtle discrimination by women faculty (Blakemore et al. 1997), the slow but steady accumulation of disadvantage over the course of women's academic careers (Valian 1999), and the flight from academia by women scientists and engineers at every step in the educational pipeline.

Today, leading universities are beginning to undertake comprehensive remedies to address these problematic attitudinal and structural issues. Prominent within the approaches being implemented are a variety of developmental, mentoring, and networking initiatives aimed at helping women S&E faculty succeed, particularly in the early and middle stages of their careers. Simultaneously, universities are undertaking a multitude of efforts in leadership development to improve the departmental and school-level microclimates surrounding women faculty in the daily conduct of their work.

In this chapter, we report specifically on the activities, challenges, and successes of a multilevel, integrated executive-coaching initiative at our university, supported through an NSF ADVANCE Institutional Transformation award. Our program is unique because we are working with both the women faculty members at all levels, and the university leaders (deans and chairs) in changing the culture of their academic units. The executive coaching experience at our university, offered to women faculty, department chairs, and deans, integrates elements of leadership development, mentoring, networking, and organizational change management.

We first define coaching and its principal purpose. We then describe the primary objectives of our coaching program, and offer an overview of the structure and the content of the coaching program, including its distinctive phases. Preliminary outcome data are also presented.

Coaching

Organizations throughout business and industry are reinvesting in their human capital through programs that enhance participants' personal and professional and help them to develop leadership. One process more frequently used in the business realm is coaching, a personalized form of improving leadership (Popper and Lipshitz 1992). The literature reports two primary purposes of coaching: to improve one's current level of individual performance and to enhance leadership effectiveness (Becket 2000; Smith 2000; Waldroop and Butler 1996).

Both coaching and mentoring are essential ingredients of a comprehensive career development strategy, although they vary in scope, duration, and focus. Traditional mentoring is a long-term relationship with a more senior colleague who provides advocacy and advice on an individual's career development, helping one learn the ropes in the chosen field and providing support for upward mobility (Kram 1985). Mentors are senior counselors who transfer knowledge, wisdom, and sponsorship (both general and disciplinary) aimed at providing access to important academic and scientific resources and networks. Effective mentoring generally relies on the development of a relatively close, mutually reinforcing, long-term, and stable relationship between the mentor and mentee.

Coaching is targeted, finite, and focused on improving current performance and thinking strategically for the long term. A coach provides an outside perspective, seeking to develop an individual's skills so the person will have the capacity to better manage his or her career and growth as a leader. Coaching tends to be more of a one-way process than mentoring, with the coach providing feedback, skill development opportunities, and guidance to the person being coached. Coaching is fundamentally focused on assisting individuals in achieving concrete goals that will enhance their learning and professional growth, as well as advancing the organization's objectives to create an environment that facilitates the success of all. Coaching in our academic context helps faculty and administrators develop skills to be more effective academic performers in the short and long run, and to implement positive change in their departments and schools. The coaches also encourage faculty and administrators to seek mentors to help them develop the long-term networks and resources necessary for accomplishment in specific academic disciplines.

In industry, many organizations are choosing to use coaching as a developmental intervention for their senior and high-potential management executives in order to bring about an organization-wide culture transformation. As Sherman and Freas (2004, 7) note, "systematically coordinating one-on-one coaching interventions that serve a larger strategic objective" fosters cultural change that benefits the entire organization. An in-depth focus group study of the state of women faculty at our university identified department chairs and school or college deans as instrumental agents in successful faculty development and university change efforts, and also found that the academic climate experienced by women is heavily influenced by the attitude and approach of these key individuals (Resource Equity Report 2003). Thus, we decided to base our process on transformation of the attitudes and behaviors of key change agents (deans and chairs) and empowerment of women faculty to proactively and collectively address the issues affecting their academic progress within their disciplinary environments.

Only by affecting attitude and behavioral changes at all levels will we realize sustainable, institutional change. Our program is not designed to "fix the women" but rather to raise awareness of gender equity issues at all levels and provide women faculty and administrators with the skills and tools to positively impact the university and academy environment. Empowering a critical mass of individuals who recognize the importance of creating more inclusive academic climates is one way to engender lasting change.

The Executive Coaching Intervention at Case Western Reserve University

The Academic Careers in Engineering and Science (ACES) project was created through the receipt of the university's NSF ADVANCE Institutional Transformation award, to catalyze faculty empowerment, leadership development, and sustainable cultural transformation in the science and engineering disciplines at our university. In 2003–4 we began this five-year university project involving the implementation of executive coaching, borrowed from business and industry and adapted for an academic environment. Specifically, this project included *(a)* executive coaching (leadership development) of individual S&E deans and chairs, *(b)* coaching of women faculty on performance, career, and leadership, and *(c)* related developmental inputs for deans, chairs, and women faculty, including opportunities for mentoring, networking, training, and development.

The ACES project offers executive coaching to administrators and chairs and women faculty of thirty-one S&E departments on our campus over the five-year period of the ADVANCE award, with the intention of institutionalizing executive coaching, mentoring, networking, training, and other developmental opportunities after the grant ends. Invitation letters are sent to each female faculty member and to chairs and deans of the participating departments, stating the overall objectives of the ACES program and the coaching initiative, and inviting them to participate in the coaching opportunity. While participation is voluntary, it is strongly encouraged as a way to further promote the goals of the ACES ADVANCE initiative, which has been championed at the highest levels of the university administration.

In the first year of the ACES project, two deans and three chairs, all male, and sixteen women faculty of four science and engineering departments received executive coaching. The ranks of the sixteen women faculty were as follows: one adjunct professor, four full-time lecturers or instructors, five assistant professors, five associate professors, and one full professor. In the current second year of the project, two deputy or associate provosts (one female and one male), two deans (both male), two associate deans (one female, one male), ten chairs (all male), two associate chairs (one of each gender), and twenty-seven women faculty of ten science and engineering departments are participating in the interventions. The ranks of the twenty-seven women faculty are as follows: one instructor, twelve assistant professors, four associate professors, and seven full professors. The female associate chair, associate dean, and deputy provost are also included in this total of twenty-seven.

The executive coaching component of the ACES intervention is staffed by professional executive coaches who have general business or organizational experience, and are experienced in providing performance and career-related advice, and guidance in developing leadership. These coaches are persons who have engaged extensively in executive coaching in corporate settings. Most were well known to the co–principal investigator responsible for the ACES coaching intervention because they had worked with her on previous corporate engagements in leadership development through the university's Executive Education Center of the School of Management. Many of the coaches held doctoral or master's degrees in organizational behavior or psychology, and some were faculty or staff members in nearby universities. Since some of the coaches were less familiar with academic than corporate settings, there were a number of initial meetings and activities, including reading assignments and discussions, that helped the coaches gain deeper knowledge of the academic work context. Regular (bimonthly) meetings of the coaches' cohort group with the ACES co–principal investigators were conducted throughout the first year of the executive coaching intervention to provide further context for the coaches and to enable confidential debriefing of experiences, successes, and challenges. Since these discussion-oriented meetings proved to be highly useful to the coaches, they were continued beyond the original test department phase into the full implementation phases of the ACES project. NSF ADVANCE resources as well as existing university expertise are utilized to support all executive coaching-related activities.

Through executive coaching, participants are helped with identifying their career and leadership vision, goals, plans, and actions. The coach gives advice, resources, and feedback on how best to accomplish the identified vision and to deal with other performance-based problems and opportunities. The duration of coaching sessions is approximately one year, with an average of six sessions for women faculty and ten sessions for chairs. For deans of the schools and colleges, the coaching intervention duration is two years for approximately ten sessions. Deans and chairs undertake a 360-degree leadership competence assessment and receive feedback about their results. The assessment is a multi-rater instrument completed by the deans and chairs themselves, their supervisors, their peers, and individuals who directly report to them. Taken together, ratings of other people provide a more complete picture of the individual's performance. As part of the ongoing coaching development, coaches meet regularly to discuss progress, challenges, and strategies for future coaching sessions. Evaluation of the professional coaching experience occurs at mid-intervention and end-intervention periods.

Targeted coaching initiatives designed to assist academic decision-makers such as deans and department chairs in understanding their roles in creating inclusive, supportive environments can help curb the leaky pipeline of faculty women in sciences and engineering. We also believe that the combined focus of limited-term coaching targeted at empowering personal and professional development together with long-term mentoring, sponsorship, networking, training, and other development inputs can help women faculty better navigate the shoals of academia. To illustrate how the coaching program is working at our university, we provide details of the objectives and stages of the process for chairs and women faculty members of each S&E department.

Executive Coaching Objectives

The primary objective of our executive coaching program is to promote academic workplace cultures characterized by equality, participation, openness, and accountability, to create a climate at the university that is conducive for faculty development, particularly for women faculty at all levels. We move toward this objective by working with deans, department chairs, and women faculty at all levels. We focus on three fundamental areas in our coaching, determined from the literature to be of paramount importance in career self-management, leadership development, and management of organizational change. First, we work to enhance the individual's self-awareness, self-confidence, and personal sense of efficacy and empowerment. Second, we assist strategic thinking about the person's career development, both for short-term effectiveness and for long-term contributions to the institution, profession, and discipline. Third, we work toward developing leadership within departments, across the university system, and among the fields of science and engineering, so that coached individuals can catalyze constructive organizational change.

The third point is worth highlighting specifically since the goal of the ACES ADVANCE initiative is to create a climate conducive to faculty achievement and success. This program is designed to work at all levels to expose and shift the underlying assumptions of the university culture and to foster an environment of inclusion that benefits the entire university community. A critical component of engendering such change is working at the leadership level.

Program Stages for Department Chairs and Deans

During the one-year period of the program, there are approximately eight to ten coaching meetings between a department chair and the coach. While the conversations and the pace of the coaching sessions are unique to each individual chair, there is a recommended series of topics to be discussed in the twelve-month period (detailed session plans follow). For each coaching session, participants are assigned preliminary and follow-up work that may require reading, personal reflection, data collection, or data analysis. Between coaching sessions, chairs meet informally as a group over lunch at regular intervals, building a community of departmental leadership.

Session 1 has two main objectives: first, to discuss the purpose and goals of ACES and the NSF ADVANCE program, and to provide an overview of the coaching initiative; and second, to discuss the chair's academic and leadership experience and current areas of interest or concern. During this meeting the coach reinforces the program objectives of ACES, particularly those related to issues of gender equity throughout the university. Chairs are asked to consider how these issues are reflected in their everyday management decisions. Questions the chairs are asked to reflect upon include the following: What are your strengths as department chair? What distinguishes your specific leadership? What have you learned as the department chair? What are your desired levels? Chairs are also asked to describe their current level of visibility and influence in the department, in the university, and in their field.

Session 2 is focused on the chairs' effectiveness as leaders. The follow-up questions from session 1 are discussed as well as definitions of effectiveness and success in leadership. The objective of this session is to help the chairs determine how to increase their impact and contributions to their departments, schools, and universities. For homework, the chairs are asked to pick a leadership role model in their field and determine what they admire about him or her.

The topic of session 3 is leadership vision and goals. The coach and chair discuss career and leadership aspirations as well as development successes and challenges. Also considered are immediate objectives as well as short- and long-term goals, individually and for the department. The concept of 360-degree feedback is explained, and the chair is encouraged to consider collecting online feedback using the Emotional Competence Inventory (ECI) (Boyatzis, Goleman, and Hay Acquisition Co. 2002) from his or her dean and other administrators and from faculty members, to assist in the process of professional development.

Session 4 is devoted to a discussion of emotionally intelligent leadership. The coach explains the concept of emotional intelligence along with the critical competencies of self-awareness, self-management, social awareness, and relationship skills. Strategies and tools for handling stress, managing conflict, taking initiative, and remaining optimistic in the face of administrative constraints are explored. The assignment for the next session is to complete online 360-degree surveys.

Session 5 is devoted to a review and discussion of the 360 Degree ECI Feedback Report. The chair's reaction to the feedback is discussed and an analysis of the overall patterns and trends in the data is undertaken in concert with the coach. As a follow-up assignment, the chair completes a self-analysis guidebook that provides a framework for deepening the chair's understanding of the feedback data.

The objective of session 6 is to discuss development planning. Based on the 360-degree feedback, professional strengths and professional development needs are identified. The chair creates a personal balance sheet of competency assets and liabilities. The coach and chair create action plans for building on strengths and addressing weaknesses linked to the chair's leadership vision, and discuss strategies for competency development. The chair receives a personal development plan template for guidance.

Session 7 focuses on gender implications for department leadership. Approaches and methods for improving the departmental climate for women faculty and students, as well as the recruitment, retention, and advancement of women, are discussed. Throughout the coaching period, the chair is encouraged to talk informally with women faculty to determine their experiences in the department; the importance of this kind of interpersonal and relationship-building behavior is reinforced during the current session. We specifically placed the discussion of the analysis and improvement of gender relations in the department after the receipt of the 360-degree feedback since the chair is then guided to consider and develop specific gender-related actions and approaches addressing issues that arise from the leadership feedback.

The topic of session 8 is increasing leadership impact and contributions as department chair. Departmental vision, goals, and culture are discussed, and the chair is encouraged to engage in strategic planning, resource allocation discussions, and ongoing communication with faculty, staff, and graduate students. The chair's personal impact and contributions as a leader in his/her school and university-wide are discussed, as well as approaches to facilitate growth in the stature of the department in its field.

Session 9 focuses on enhancing the chair's interpersonal skills. Topics include effective negotiation with higher administration and other funding sources, and learning to interact with people of different styles and personality types. Chairs consider their default interpersonal styles and approaches, especially in times of stress or conflict, and discuss how to expand the behavioral repertoire to be more effective. Specific strategies and techniques for dealing with difficult faculty and situations may also be discussed.

Session 10 deals with the closure of the coaching relationship. The coach encourages the chair to revisit insights gained from the program, continue to clarify and implement his or her leadership development plan, practice new departmental change-related behaviors, and stay focused on the goal of creating and supporting an inclusive and empowering department climate.

Case Study of the Departmental Chair Coaching Process

To illustrate the impact of the coaching program with a member of the university leadership team, we present a case study of Matthew Price (fictional name), who made significant changes in his departmental leadership as a result of the coaching initiative. Matthew had an established reputation in his field of science. He enjoyed his research and continued to be a productive contributor in his discipline. Two years earlier he had reluctantly assumed the position of chair because of his seniority. The department was of moderate size, twenty members, with an inordinate number of very junior faculty and very senior faculty. There were two women who had not yet reached tenure and one senior woman with tenure.

As chair, Matthew was struggling with a number of issues. He wanted to improve the department's stature and to establish some clear areas of research distinction for his department. He believed that disciplinary barriers were polarizing the department. The generational differences in the department created additional conflict. He aspired to continue his research while attending to a new strategic direction for the department.

During his first meeting with his coach, Matthew wondered what would enhance the possibilities for recruiting outstanding female faculty members. He fully supported the university's objectives to improve the environment for female faculty, yet was uncertain what might be done to accomplish this end. When asked what his expectations were for the coaching relationship, he offered that he wanted to learn about the impact of his behavior upon the faculty and upon the female faculty in particular. He believed that his distinctive strength was his fairness, while an area for improvement was his decision-making style.

After Matthew had completed a 360-degree feedback process, his coach met with him to review the results. In addition to strengths identified in his self-assessment, this feedback reported that other strong areas included his teamwork and collaboration, his optimism, and his adaptability. Areas for improvement were his conflict management and his inspirational leadership skills. Matthew was open to the feedback and was eager to begin working on these competencies in order to develop as a leader. He spoke with his coach about strategies for changing his behavior. For example, while Matthew had expressed reluctance to continue as department chair during his first meeting with his coach, he became more interested in leading his department through their transitional time. His coach and he discussed his vision and his goals for the department, and why they were important to him. He then went to work to pursue his vision and his goals. One strategy Matthew undertook in order to demonstrate inspirational leadership was to pursue opportunities to articulate his vision for the department with his colleagues in both group meetings and individual conversations. In the past Matthew had been reluctant to do so because of his overreliance on a democratic leadership style. He also actively directed an effort to develop a strategic plan for the department as another way to display his inspirational leadership.

As the coaching relationship came to an end, Matthew told his coach that he had already witnessed his improved effectiveness as a leader. He was working on his conflict management skills and offered some stories to illustrate his success to date. He was leading his department through their strategic planning and believed that progress was being made on that front. He felt that he had improved his relationships with his faculty members and was paying special attention to the thoughts and feelings of his female faculty members. Anecdotal reports (during informal conversations with the principal investigators of the award) from all three of his women faculty indicated that they felt their chair had changed substantially over the coaching period, primarily to increase his sensitivity to, and awareness of, the issues and challenges facing women in academia, and also with regard to his effectiveness as the department's leader. Toward the conclusion of the coaching activities, Matthew was excited about the future possibilities for himself and for his department.

Program Stages for Women Faculty Members

Another critical constituent group for the coaching initiative is women faculty. In addition to working at the level of university leadership, it is also important to work with individuals, assisting women faculty in developing the skills and resources necessary to achieve academically in the university system. Approximately six coaching sessions occur between the individual women and their respective coaches in the course of one year. The program is very similar to the program for department chairs, but the recommended topics for discussion are targeted at assisting women faculty in leveraging their professional contributions for maximum impact.

The first coaching session is intended to introduce the program goals and review the career history and highlights of the faculty member as well as current interests or concerns. Women faculty members are asked to assess their distinctive strengths and contributions as scholars and professionals. Performance issues or areas of opportunity for development over a woman faculty member's career are discussed, including performance, scholarship, work-life integration, mentorship, and inclusion.

The second meeting centers on the topics of professional excellence and academic success and what might increase the member's influence and contributions in her department and in her field. Participants are asked to reflect on the meaning of academic success and effectiveness, as well as on the methods by which they may gain greater professional visibility and reputation. Women faculty are encouraged to consider who their role models and mentors are, and how they can widen their network of connections and influence in their academic sphere. At this session, each participant is asked to begin the process of creating a mentoring committee, consisting of a senior departmental colleague, a university colleague, and a senior visible scholar in her specific discipline or academic area, that will meet periodically over the next two years to assist in her academic development and advancement.

Immediate, mid-term, and long-term career aspirations, challenges, and goals are discussed in the third coaching meeting. In the interim between the second and third coaching meetings, participants observe a role model's style, behavior, presence, and influence, and interview the role model about his or her development, choices, advice, and so on. Based on their insights from this interview, participants identify a personal vision of academic and professional excellence, which in future sessions they will translate into more specific goals and actions.

At the fourth meeting, women faculty members are introduced to topics of interpersonal skills, emotional intelligence, and strategies for enhancing interpersonal competencies. Unlike the department chairs, women faculty are not routinely asked to participate in the 360-degree feedback process, with the exception of senior, tenured women, if they so choose. Since many of the environments in which women faculty are embedded are "chilly climates" (Sandler and Hall 1986), and since the majority of women faculty being coached are junior professors, the coaches are careful to invite 360-degree feedback only if participants seek out this information or are placed in leadership positions requiring feedback on their leadership competencies.

The remaining two meetings between women faculty and their coaches encompass the continued identification of goals, professional strengths and areas for growth, a personal plan, and ideas for sustaining career development. Discussion is also focused on enhancing the faculty member's interpersonal skills, especially for handling difficult colleagues and situations. Finally, the coaching relationship concludes with the discussion of plans to sustain individual performance and contributions, and overall career development.

Like the process for the chairs, for each coaching session women faculty are assigned preliminary and follow-up work that may require reading, personal reflection, data collection, or data analysis. Between coaching sessions, participants meet informally over lunch at regular intervals, and attend formal educational events such as workshops on success in academic careers, mentoring, and negotiating.

Preliminary Evaluations

The results of the executive coaching initiative have been most encouraging. Participants' ratings of their specific coaching experiences were extremely positive. Overall, in the first year of the program, effectiveness of the coaching process was rated at 4.46 out of 5 at the mid-intervention period ($N = 13$), and final coaching evaluations for first year participants averaged 4.56 out of 5 ($N = 9$). In the second year, mid-intervention ratings of overall coaching effectiveness were 4.19 out of 5 ($N = 21$). While these are small numbers, the positive evaluations are encouraging.

Anecdotal descriptions of the effectiveness of executive coaching from women faculty in the first year of the program indicate that they have used their broadened networks to obtain valuable career supports, including guidance on tenure package write-ups, reviews of their curriculum vitae, advice about teaching, visits to external universities, invitations to panel discussions at conferences, and means to more effectively manage their labs, postdocs, and graduate students. Comments from women faculty at different academic levels included the following:

> My coach has been very effective in directing me to communicate with my colleagues and my chair.

> [My coach was] helpful in getting me to set priorities and to stop procrastinating about accomplishing an important work goal. Also my coach held me accountable to move ahead with my mentoring committee meeting which turned out to be very helpful.

> Initially I felt the coaching was a bit of a waste of my time. The sessions seemed very unfocused and "chatty" rather than specifically helpful. However they took a dramatic turn for the better as they progressed, and I found the overall experience very positive. My coach helped me to clarify my career and life needs and goals, and helped me recognize when I was living my values and when my actions were torpedoing them. This reduced my stress level considerably and made me more effective.

> Overall this was a good experience. The one downside is that the sessions did raise issues about gender inequity that I have managed to suppress over the years in order to survive. So in a way this experience has opened a "Pandora's box" and I am more aware and angered by these inequities. It is now a matter of finding some balance in identifying some issues that can be realistically addressed and putting the others back in the box. I think it is difficult to achieve success if you only focus on the negative aspects of the culture. Admittedly sometimes it is easier if you pretend they don't exist.

The evidence thus suggests that women are finding the coaching practical, applicable, and relevant in creating a strategy for career development. Executive coaching is effective in empowering women faculty to gain control over their academic careers and performance.

More importantly, we also conducted an analysis of pre- and postdata on attitudes and knowledge about career and professional success. Women faculty rated themselves higher on a variety of factors after executive coaching was implemented for a year.[1] Postintervention ratings increased for items such as clarity about career direction, articulation of career goals to others, exercise of initiative toward attaining career goals, mentoring sought from outside the department, influence exerted and success experienced in the department and the field, and collegiality and leadership exercised in the department (see table 1). Interestingly, postintervention ratings dropped for mentoring received and given by women faculty within their departments, as did ratings of career progress and academic or scholarly contributions. These latter ratings suggest that for some women faculty, the coaching processes may have provided a more realistic view of their career progress, successes, and contributions; women faculty appear to have received newer external feedback and engaged in different internal reflections about their success, prospects, and career actions through the executive coaching processes.

Conclusions

In this chapter we have described a comprehensive program intended to advance the status of women faculty in the science and engineering disciplines in a university setting through individualized leadership development of key change agents (deans and chairs) and individualized empowerment of women faculty to take control of their academic careers. The program is unique for two specific reasons. First, the executive coaching initiative encompasses multiple professional developmental activities (i.e., mentoring, networking, negotiating, development planning, leadership development, strategic thinking), and second, it is targeted at multiple levels within the university hierarchy (i.e., junior and senior women faculty, department chairs, and deans). The executive coaching intervention facilitates empowerment aimed at career development and improvement of academic performance and faculty leadership. Preliminary evaluations indicate that this kind of developmental program has the potential to positively affect how women faculty and leaders of schools and departments approach change efforts in the university.

The implications for women faculty from our executive coaching initiative are already evident at this early stage in the process of institutional transformation. Women faculty members are becoming more aware of the need to be strategic in planning for their long-term academic success. In addition to systematic planning for career development, they recognize the critical importance of establishing a network of individuals in their disciplines who can help them gain access to resources necessary for academic achievement. They feel more in con-

TABLE 1. Pre- and Post-outcome Evaluations for Women Faculty in the Sciences and Engineering Departments, Year 1

Item	Preintervention (N = 12) Mean	Postintervention (N = 8) Mean
1. Are clear about career direction and goals in the next 5 years	3.25	3.38
2. Are able to clearly articulate your career direction and goals to others	3.25	3.38
3. Have exercised initiative towards attaining your career goals	3.33	3.63
4. Have taken proactive steps to increase your scholarly visibility	2.33	3.50
5. Are clear about the role of a mentor	3.58	3.13
6. Have actively sought mentoring from within your department	3.33	2.75
7. Have actively sought mentoring from outside your department	2.42	3.00
8. Mentor other colleagues in your department	2.42	2.25
9. Mentor students/postdocs in your department	3.67	3.63
10. Exert influence in your department	2.17	2.75
11. Exert influence in your discipline/field	2.08	2.50
12. Feel successful in your department	2.33	2.63
13. Feel successful in your discipline/field	2.58	2.63
14. Feel a sense of control over your work and environment	2.50	2.63
15. Are able to balance multiple priorities and effectively use your time	3.00	3.00
16. Your current career opportunities	2.67	2.50
17. Your career progress to date	2.92	2.63
18. Your overall academic/scholarly contributions	3.00	2.63
19. The colleagueship you provide in your department	2.91	3.29
20. The leadership you provide in your department	2.64	3.00
21. Your likely career success in the next 5 years	2.70	2.75

Note: Items were rated on a four-point scale, 1 = Not at all, 2 = To some extent, 3 = To a moderate extent, 4 = To a great extent.

trol of their career outcomes, and are taking steps to increase their overall visibility, contributions, and impact, not just in their departments and university, but also in their larger academic fields.

Our coaching interventions for department chairs and deans have resulted in greater awareness of the importance of leadership development training within the academic hierarchy, and enhanced knowledge of the subtle institutional biases against women that may exist in traditional academic systems. For example, one of the chairs mentioned in his evaluation of the coaching program, "I am a more aware and communicative leader as a result of this coaching." Another chair strongly advocates that all male faculty should be required to read *Why So Slow* by Virginia Valian, a book that explains the existence of gender schemas and lays out the systemic prevalence of gender bias. Other chairs have indicated that they would like further opportunities for peer-driven leadership development; subsequently our project has organized more frequent networking opportunities for chairs, and a more formalized leadership development program for chairs through the provost's office. Thus, the first step toward institutional change is the recognition of problems that need to be addressed. Working with the system's leadership (chairs and deans) to highlight existing issues of gender inequity and to develop strategic plans to work at both the individual and the systems level is critical to engendering lasting organizational change.

Fully measuring the degree to which the multilevel coaching initiative results in organizational change will require longitudinal research. Quantitative measurements and in-depth qualitative studies to gauge numeric changes and behavioral shifts in individuals as well as to determine modifications in the university systems can assist our understanding of the coaching program's impact on both individual learning and organizational transformation.

Organizations are systems of patterned behaviors, and cultural transformation is a slow process. This is particularly true in traditional hierarchical organizations such as universities. Profound and sustained change occurs when there are shifts in the norms, mental models, and shared assumptions, leading to a transformation in the systems and the practices of the departments and schools within the university. Organizations that sponsor developmental activities such as those employed in our executive coaching initiative may witness improved retention rates (Higgins and Thomas 2001) and enhanced organizational success (Tannenbaum 1997). The multilevel integrated coaching program described in this chapter is designed to support a shift in the perspectives and mind-sets of faculty members, department chairs, and deans. We believe changes in the gender schemas of a critical mass of university faculty and administrators can result in a cultural revolution that improves the entire university.

NOTES

This study was supported by an NSF ADVANCE Institutional Transformation Grant, SBE-0245054, start date 9/1/2003, end date 8/31/2008.

1. We could not utilize tests of significance for these pre- and postdata analyses since the sample sizes were small and it was not possible to match responses by the same person over the two data collection points.

REFERENCES

Becket, M. 2000. Coach class, top class. *Daily Telegraph,* October 19 .

Blakemore, J. E. O., J. Y. Switzer, J. A. DiLorio, and D.L. Fairchild. 1997. Exploring the campus climate. In *Subtle sexism: Current practice and prospects for change,* ed. N. J. Benokraitis, 54–71. Thousand Oaks, CA: Sage.

Boyatzis, R. E., D. Goleman, and Hay Acquisition Co. 2002. Emotional competence inventory. Boston, MA: Hay Group.

Higgins, M. C., and D. A. Thomas. 2001. Constellations and careers: Toward understanding the effects of multiple developmental relationships. *Journal of Organizational Behavior* 22:223–47.

Kram, K. E. 1985. *Mentoring at work.* Glenview, IL: Scott Foresman.

Popper, M., and R. Lipshitz. 1992. Coaching on leadership. *Leadership and Organizational Development Journal* 13 (7): 15–18.

Resource Equity Report. 2003. Resource equity at Case Western Reserve University: Results of faculty focus groups. http://www.cwru.edu/menu/president/resourcequity.doc.

Sandler, B. R., and R. M. Hall. 1986. *The campus climate revisited: Chilly for women faculty, administrators, and graduate students.* Washington, DC: Association of American Colleges and Universities.

Sherman, S., and A. Freas. 2004. The wild west of executive coaching. *Harvard Business Review* 82 (11): 82–89.

Smith, J. 2000. Coaching and mentoring. *Stress News* 12 (2): 12–14.

Tannenbaum, S. I. 1997. Enhancing continuous learning: Diagnostic findings from multiple companies. *Human Resource Management* 36:437–52.

Waldroop, J., and T. Butler. 1996. The executive as coach. *Harvard Business Review* 74 (6): 111–17.

Valian, V. 1999. *Why so slow? The advancement of women.* Cambridge: MIT Press.

THIRTEEN

Interactive Theater

RAISING ISSUES ABOUT THE CLIMATE
WITH SCIENCE FACULTY

Danielle LaVaque-Manty, Jeffrey Steiger, & Abigail J. Stewart

INTERACTIVE THEATER can be used to raise political consciousness, provide therapy, even develop legislation (Boal 1997). In a recent pilot study, Chesler and Chesler (2005) found it an effective tool for building community among female faculty in engineering, and Brown and Gillespie have used it to confront what they call (following ethicist Andrew Jameton) "moral distress" in the university—situations in which "we believe we know the right thing to do, [but] feel constrained from doing it because of stultifying demands or practices over which we have little control" (1999, 36).

At the University of Michigan, we have found that interactive theater techniques can offer a surprisingly effective way to raise issues about the climate with science and engineering faculty. Sketches illustrating how faculty interactions shape and reflect the climate—portrayals of discussions of job candidates in department meetings, efforts of senior faculty to advise and mentor junior faculty, and committee meetings evaluating tenure candidates—have been used with a range of audiences to stimulate actor-audience interactions that raise key issues about how gender, rank, ethnicity, and other aspects of power relations influence the climate and faculty morale.

Imagine that you are a fly on the wall at a department's faculty meeting, observing conversations about the relative merit of two candidates for an open faculty position: one is an innovative junior woman just finishing a postdoc, the other a man on the verge of tenure, working very successfully in the mainstream of his discipline. The only woman faculty member at the table is suffering repeated interruptions of her well-articulated arguments on behalf of the female candidate while receiving support only from one junior male colleague. Many aspects of the discussion are familiar, some of them perhaps painfully so.

Imagine, now, that you become visible to the people at the table and they invite you to ask them why they did what they did and said what they said during the course of the meeting—a chance, in other words, to bring into the open the personal motivations, group dynamics, and political subtexts that usually remain unexplored and unacknowledged during conflicts among faculty who must work together on a daily basis. Imagine telling the department chair that he isn't doing his job very well when he allows his male colleagues to keep interrupting their female peer. Imagine that you do this without putting your own career or anyone else's at risk.

Interactive theater can simulate such an experience; and, intriguingly, the fact that the faculty meeting is neither "real" nor a traditional dramatic performance that can be passively witnessed may be of great advantage; the audience is asked to be aware of itself observing and participating in a staged conversation for the purpose of thinking about problems that are difficult to engage in the abstract.

In sketches presented by the Center for Research on Learning and Teaching (CRLT) on behalf of ADVANCE at the University of Michigan, brief but complex scenes encompassing common faculty dynamics are enacted for audiences of faculty and administrators. Following the scene, a facilitator invites the audience to ask questions of the actors, who remain in character. At first, the actors respond as they would if they were still in the presence of their colleagues, but "time-outs," during which an actor is invited to respond as if his or her character is the only one at the table, allow for more frankness.

Audience members often disagree with one another's interpretations of the scene. Some regard the climate in the hypothetical department as toxic and sexist, while others may assert that the female faculty member simply needs to be more aggressive. The facilitator keeps the dialogue moving, with certain directions in mind, and may conclude the discussion with some reframing to ensure that audience members have clear ideas to think about later. Audience members are also given folders containing relevant reading material to take home with them.

An example may help illustrate how the process works. Consider the following exchange from the faculty meeting sketch:

MARLENE: Yes . . . there are very different reasons for hiring people of different talents at different places in their career trajectory. Yes, we need to acknowledge that they are in different places. And it is because we are acknowledging this that we need to really think about—

FRANK: *(Speaking over* MARLENE*)* Well, Professor Young is at a place in his career that does make him much more sought after and much more influential. . . . The prestige he will bring to this department is unparalleled. We need to think about how our department will be perceived . . .

(MARLENE *looks at* TERRANCE *as she and* FRANK *overlap. He does nothing.*)

MARLENE: *(To* FRANK*)* Excuse me, I am not done speaking. (Steiger 2004, 5)

In response to this scene, one audience member may ask Marlene (who, the audience knows, is tenured) whether she has ever considered looking for a position elsewhere, while another might ask why she doesn't simply speak up more. Alternatively, someone might ask Frank why he interrupts Marlene so much, or ask Terrance why he doesn't direct the discussion in a way that allows everyone to be fully heard.

If the conversation portrayed in the sketch reveals as many complex social dynamics, power relationships, and apparently "individual" concerns as are embedded in an actual faculty meeting in the real world, it presents a web of problems its audience cannot easily solve or dismiss. Engaging in conversation with actors playing the roles of faculty members may sound childish to many faculty at first, but those who might quickly dismiss certain characters as buffoons or exaggerations are forced to think differently when they address those characters directly and are confronted with intelligent and complex justifications for their behavior. For example, an apparently "unassertive" Marlene may rebuke an audience member who accuses her of passivity and challenge audience perceptions that she just needs to be "more assertive" in order to solve her own problems. The success of the theatrical interaction depends, like most academic exchanges, on argument.

The CRLT Theatre Program

In addition to the faculty meeting sketch described above, the CRLT Players have developed a sketch about faculty mentoring and another about tenure evaluations on behalf of ADVANCE, but work related to ADVANCE and its goals is only part of the group's repertoire. The CRLT Theatre Program has historically focused on classroom dynamics, and most of its sketches were designed to help instructors, whether faculty or graduate students, improve their teaching. Those sketches focus on gender, race, and disability in the classroom, among other topics.

Established in 1962, CRLT was the first teaching center of its kind in the United States. The CRLT Theatre Program began in 2000 with just one sketch. At that time, Jeffrey Steiger served as the director in a full-time position. Its initial budget, contributed by the deans of the College of Engineering and the College of Literature, Science and the Arts (LSA), amounted to forty thousand dollars per year. Today, CRLT Theatre employs not only Steiger but also an assistant director on a full-time basis. The troupe's repertoire includes fourteen sketches and its yearly budget has grown to $250,000.

Because collaboration with the CRLT Theatre Program was part of Michigan's NSF grant proposal, UM's ADVANCE Institutional Transformation project commissioned a set of sketches from CRLT immediately upon receiving its grant in January 2002, and committed funds to support the development and performance of three theater sketches over a five-year period. To date, the theater program has developed four sketches for ADVANCE. The first was discarded, for reasons that will be discussed below. The second and third are being performed regularly, and the fourth was rolled out for regular performances during fall 2005.

The sketches CRLT has developed for ADVANCE have been based on two kinds of research: written academic work and experiential role-play. While ADVANCE and CRLT's own research staff are able to provide Jeffrey Steiger with studies on gender, science, and the academic climate for women faculty, this sort of data serves more as a source for fact checking and revision than as a well of inspiration during the creation of a sketch.

The first sketch Steiger developed, which was later dropped, portrayed a woman faculty member's difficulties in establishing authority in an all-male (or nearly all-male) classroom. This sketch was similar to other sketches the CRLT Players were already performing in that it focused on the teacher-student relationship. It drew from a campus climate study, and particularly from ADVANCE interviews with women faculty, for its understanding of the problems women might face in such a situation. This sketch also used a specific interactive technique drawn from Forum Theatre (Boal 1997)—one that CRLT often employs—of inviting a member of the audience to replace the actor playing the part of the woman faculty member on stage and play the role differently, in the hope of generating a better outcome.

All CRLT sketches undergo a preview process during which the actors and director receive feedback from knowledgeable audiences. Does the language in the sketch ring true? Are the characters persuasive? Are the facts correct? Do the actors, when they interact with the audience, give appropriate responses to various questions?

Women who previewed the first sketch felt victimized by the way it worked; it was, in effect, set up to "blame" the woman faculty member's character for the difficult dynamics in her classroom and to invite audience members to feel superior to her as they "corrected" her approach during their turn on stage. This revealed a potential drawback to the Forum Theatre approach; most of the later sketches have involved direct interaction between all of the actors and the audience rather than replacement of actors with audience members.

Script Development

Steiger discovered, as he began to develop a new sketch—the faculty meeting, which includes no student characters—that while he and the other actors understood how faculty-student relationships worked because all of them had been (or were still) students, they did not know what life was like for faculty outside the classroom. They did not grasp the spoken and unspoken rules of academia. They did not understand the basic facts of how departments function.

Further, in the classroom sketches, the actors work to understand their characters as individuals in the classroom context, pinballing off of, and connecting to, one another as students involved in a temporary relationship within a climate created by the instructor. In contrast, in the ADVANCE sketches, the performers need to understand the long-term relationships among characters behaving in accord or opposition within the "whole" of not only the department, but the entire university. Each character, connected to both micro and macro levels of the institution, is a personality operating within a hierarchy.

Steiger thus developed a new method for creating the ADVANCE sketches. He now begins by meeting with a group of faculty who can talk to him about what their world is like and help him identify a scene that will resonate with a faculty audience. He then stages a role-play involving actors who know the environment that is the focus of the sketch. (The theater program employs professionals with formal theater training, students, research staff, and others as actors in the troupe.) Not everyone involved in the role-play has to be part of the world portrayed, as long as some of the participants are familiar with the norms and language typical of that setting. The initial role-play used to develop a sketch on faculty mentoring, for example, involved two members of the ADVANCE staff. If the participants have seen moments that are representative of the interactions the sketch will ultimately portray, they will naturally enact the subtleties of their experience of those situations and bring them to life in ways that are both intentional and unintentional. (An additional benefit to using faculty in the role-play is that those involved in the process become allies of the sketch and the program in general. This is one way the CRLT Theatre Program creates a network of supporters for its work.)

Armed with what he has learned from observing the role-play, Steiger is able to give the acting troupe an overview of the story they will portray and the culture in which it takes place. Actors are assigned parts in a script created from a transcript of the original role-play. Steiger also asks his actors to engage in exercises or workshops that Anne Bogart (2001) might call *source work;* a series of activities done at the beginning of the rehearsal process to connect intellectually and emotionally with the script. Actors presenting a scenario on gender dynamics in a science classroom, for example, may participate in an experiential exercise that instructs them to list adjectives that are most associated with or best describe the traditional roles of a man or a woman. Players may be asked to share experiences they had growing up that carried their first "lesson" regarding gender roles.

Actors, like everyone else, have biases and limitations based on their own particular backgrounds and experiences. By engaging in source work with their fellow actors, they become more able to view their own characters in a more three-dimensional way, rather than through the lenses of their own individual presumptions and predilections. They are also better able to understand the forces that prevent change or empathy in an individual character by exploring their own resistance and presumptions, and through this process, better able to push an audience that may have points of view similar to those the actor held before engaging in the exercises. Doing the source work also improves relationships among the actors—a necessarily diverse group—within the theater troupe itself. Before Steiger began using source work, his troupe suffered a much higher rate of actor turnover, leaving him repeatedly with groups that were all white. Using source work has reduced turnover and enabled the troupe to retain actors of color.

Simply including an actor of color as a cast member changes an audience's reaction to the faculty meeting sketch in more and less subtle ways. When one of the male faculty members in the sketch is African American, for example, race tends to arise as a topic of conversation in the audience interaction, while race is unlikely to be discussed when the same character is played by a white actor. This occurs despite the fact that the scripted lines are the same no matter who is cast in the role. The frequent assumption on the part of a mostly white audience is that the character of color is "selling out" or shirking a race-based responsibility. This perception can initiate an enlightening and contentious discussion regarding assumptions about race, power, and culpability.

The Role of Feedback in Sketch Development and Promotion

Once the characters have been developed and the parts learned, the troupe is ready to collect feedback from carefully selected critics. CRLT staff, ADVANCE staff, and members of a faculty advisory committee now called Strategies and Tactics for Recruiting to Improve Diversity and Excellence (STRIDE), another ADVANCE intervention devoted to improving recruitment and hiring practices in science and engineering at the UM, serve as test audiences. As discussed in another chapter in this volume, the members of STRIDE are all well-respected senior men and women in science and engineering fields who have studied social science literature on gender in academe and who have become activists on behalf of women science and engineering faculty. Because STRIDE members not only thoroughly understand the goals of ADVANCE with respect to these sketches, but also have a well-developed sense of how their science and engineering colleagues are likely to respond to various aspects of the performances, their feedback is particularly valuable during these preview sessions. STRIDE members attend performances of all ADVANCE sketches, if possible, to ask useful questions if discussion is slow in getting started and keep it moving should it lag.

During a preview, the facilitator asks the audience to interact with the characters as a real audience would and to ask specific questions that might be difficult for the performers to answer. This conversation is itself a kind of training for the actors, who discover things they still don't know about their characters' lives when they run up against questions they can't answer. At the end of the preview, the facilitator asks the audience to step back and give feedback about how the sketch might be improved. In addition to providing necessary information to the CRLT Players, this process helps create faculty buy-in, because faculty who see a sketch at this early stage feel like consultants involved in its creation and take ownership of it.

Another key preview audience for the sketches is the Network to Advance Women Scientists and Engineers—an informal network, supported by ADVANCE, that includes all tenured and tenure-track women science and engineering faculty at Michigan. Typically, the network is invited to an informal dinner at which the sketch is performed. Again, this is a key audience for the sketch; it best represents those the sketch is ultimately designed to help, and it is able to point out aspects of the performance that might unintentionally portray women faculty in ways that could be counterproductive or put them at risk, as in the case of the discarded teaching sketch.

All ADVANCE sketches were also previewed by the Academic Program Group (APG), which includes the provost, associate provosts, and deans. University of Michigan president Mary Sue Coleman also attends some APG meetings and came to the one at which the faculty meeting sketch was performed. The sketch was well received at this presentation, which both affirmed the CRLT Players' sense that their performance was convincing and enabled ADVANCE principal investigator (PI) Abby Stewart and CRLT director Connie Cook to promote the sketch by referring later to its positive reception by the University's provost and president. President Coleman, in fact, was quite enthusiastic about the sketch, and has since promoted it at national meetings of educational leaders. The APG preview, then, legitimized the sketch in multiple ways. Finally, during the period in which it was developed and rolled out, the faculty meeting sketch was also previewed for LSA's dean, along with his entire staff; as a result he and the ADVANCE and CRLT staffs strategized about how to use the sketch most effectively in his college.

Preview audiences are also asked for advice about the composition of future audiences. For example, the faculty meeting sketch was deemed to be potentially explosive if performed in a department because of the likelihood that the issues portrayed on stage could be mapped onto real and ongoing conflicts in the department. Thus, with one exception to date, audiences for this sketch have been drawn from multiple departments. LSA began by having the sketch performed for all of the department chairs in the college, with the idea that chairs might be willing to help promote the sketch to their faculty. This performance raised new issues. First, the importance of setting was underscored by the fact that the sketch was presented (as a result of building renovations) in a room that was uncomfortably small for the group. Even more importantly, despite the facilitator's valiant efforts to engage the chairs in a fruitful discussion of the group dynamics in the meeting, the discussion remained focused on procedural and mechanical issues. In retrospect it seemed clear that the "real" chairs were not willing to point out the crucial inadequacies of the chair's performance in the sketch situation.

Strategies for Framing the Sketch

The experience with the chairs underscored the importance of framing of the sketch, and providing a context in which the discussion could be relaxed and fruitful. The dean of LSA decided personally to invite faculty in the LSA science departments to dinners at which they would view the sketch. Three dinners were held, and forty senior faculty members from natural science departments were invited to each one. Faculty were seated at tables that ensured mixing of faculty across departments, and an effort was made to include at least a couple of women faculty at each table, as well as at least one person familiar with the sketch and associated with ADVANCE efforts within the University.

At the beginning of each dinner, the dean pointed out that tables had been deliberately mixed because faculty so seldom meet those in other departments, and he asked each person to stand up and introduce him- or herself. He framed the dinner as an effort to create more community, and the performance as an effort to pay more attention to the community's climate. He also made concluding comments at the end of each evening, pointing out dynamics he noticed in the sketch that he found particularly illuminating with respect to issues he had encountered in real life. The dean's presence at these events and active engagement with the sketch was extremely validating. LSA faculty responded to these performances with thoughtful and positive comments, often focusing on how convincing the portrayal had been and mentioning issues they had continued to think about afterward, like whether they themselves were truly "equal opportunity interrupters" or interrupted women more often than men.

In our experience, the sketch has been less well received in settings in which unit leaders did not attend the performances themselves, or when the sketch was performed in a less hospitable setting, absent a meal and an opportunity to interact with colleagues. In such cases, faculty can view their attendance as simply fulfilling another onerous work obligation, and the impact of the sketch is reduced. It should be noted, though, that we have learned that it is important to be open to experiment with the sketches.

Despite the fact that there were concerns about showing the sketch within a single department, one LSA science department chair requested a performance for his department. He believed it might offer his faculty an opportunity to critique both their own group dynamics and his behavior as chair, and he was interested in encouraging that kind of reflection. In fact, when the audience was invited to interact with the actors, he asked the first question, and voiced clear criticism of the chair. His active questioning enabled the women assistant professors who attended the performance to ask many questions of their own. While not every member of the department attended, those who came were very engaged and continued to discuss the issues raised long afterward. The chair also reported later that he received useful feedback from his faculty over the subsequent two weeks. This was perhaps an unusual case, because this particular chair was interested in identifying and addressing his own limitations. This experience also underscored the importance of the form of participation engaged in by the visible leaders at these presentations.

Other ADVANCE Sketches

Two additional sketches are still in a process of being "deployed" on campus, though they are at different stages. The second sketch portrays a male senior faculty member attempting to mentor a junior woman. Understandably busy, the senior faculty member doesn't really clear much time in his day to talk to the junior woman, nor does he read her work or even her vita very carefully before she arrives in his office, despite the fact that she sends it to him well in advance of the meeting. The advice he gives her, though well intentioned, is entirely discouraging, and the meeting is interrupted by a junior male faculty member with whom the senior male clearly has a more cordial relationship.

After very positive receptions from the preview audiences, and a general sense that this sketch evoked much less defensiveness among faculty than the faculty meeting sketch had, an effort was made to collaborate with the LSA dean's office in presenting this sketch as part of a multiyear effort to improve mentoring practices in the College. The sketch was presented in the successful dinner format at multiple dinners for all department chairs in the College. These discussions were intended to lay some groundwork for a more explicit consideration of mentoring policies and practices in the departments. Chairs were provided with copies of a new *Faculty Advising Faculty Handbook* that had been developed by Professor Pamela Smock of Sociology and ADVANCE staff member Robin Stephenson, and they were encouraged to share the handbook with senior and junior faculty members in their departments.

A follow-up workshop was held at which the dean, ADVANCE PI Stewart, and Professor Smock gave presentations on the contents of the handbook, and participants worked in small groups to develop templates for departmental mentoring plans that would maximize good mentoring outcomes and minimize bad ones. This required extensive discussion of what would count as good and bad outcomes, so the participants generated a list that allowed the workshop leaders to develop a template to send back to everyone who attended. Departments were then asked to use the template to generate more specific departmental mentoring plans and given a year to do so. All departments have at this time submitted mentoring plans, and during the upcoming academic year departments will be encouraged to present the mentoring sketch to their faculty as part of an effort to increase awareness of effective mentoring practices. The mentoring sketch, then, was presented in the context of a larger project that gave those who saw it reason to take it seriously and make use of what they learned soon afterward. This kind of framing is critical if the sketches are to be absorbed and used by those who see them, rather than merely watched and forgotten.

Finally, the tenure evaluation sketch was developed to address problems of evaluation bias in the tenure process. Because the faculty meeting sketch and the mentoring sketch had already been so well received, there was widespread agreement among university constituents, including the provost, that the CRLT Theatre Program was the appropriate tool to use, and Jeffrey Steiger readily agreed to develop a script. In order to do so, he asked senior faculty to enact a role-play of a tenure discussion for him, which he observed to gain a sense of how the process works. Members of STRIDE and other supportive senior faculty performed the role-play, and afterward, Steiger asked them what *hadn't* happened in the role-play that usually happens in tenure discussions. All of them agreed that nobody had taken the part of the "bean counter," the person who always wants to tally up numbers of publications, status of publication venues, and citation rates in order to make a decision. Bean counting was thus integrated into the sketch. Steiger also asked follow-up questions about differences in practices between different colleges and at different levels of review, and received extremely detailed answers that worked their way into his script.

At this writing, this sketch has been shown to preview audiences and the rollout has begun; it has been performed for the Academic Program Group, and in two performances for key tenure decision-makers in LSA: the Executive Committee, members of the three divisional review committees, and department chairs. In this way it quickly reached a large number of the people involved in tenure decisions. Subsequent performances will be offered to groups in the College of Engineering, as well as people on department-level tenure committees, perhaps again accompanied by dinners, in order to prompt thinking about relevant issues before this year's tenure cases come up for evaluation. At all presentations a handful of journal articles addressing issues of gender bias in evaluation processes will be distributed.

Intriguingly, this sketch invites the kind of audience participation that was unhelpful in the discarded teaching sketch, but with a twist: rather than replacing any of the faculty members at the table, audience members are invited to add themselves to the table and intervene in the discussion. They are thus invited (in small groups) to think of ways to redirect the conversation without having to decide that any particular person already at the table is responsible for its failings. In the process, they are given an opportunity to practice ways of intervening in a tenure discussion that has gone awry. This strategy helps mobilize audience members' awareness that their actions (and inaction) matter in these situations, while giving them an opportunity to work with the group on identifying strategic interventions that might be effective.

Evaluating the Sketch

Evaluation of the sketches serves at least three goals: providing feedback to the theater program; offering assessments of, and justification for, the theater program; and offering assessments of and justification for the use of theater for purposes of institutional transformation. Both CRLT and

ADVANCE collect survey data regarding audience responses to each performance. This kind of aggregate data can provide a sense of the immediate impact of a performance—and assessments of performances of the faculty meeting and mentoring sketches to date show that the sketches are generally well received and thought-provoking.

We have quantitative ratings of sixteen performances of the faculty meeting sketch and seven of the faculty advising faculty sketch. Overall, 519 individuals, of whom 322 were from UM, rated the faculty meeting sketch, and 276, of whom 206 were from UM, rated the faculty advising faculty sketch. About half of the audience rated the sketches (53% of UM audiences for faculty meeting and 46% for faculty advising). Three items invite audiences to rate the usefulness of the issues and topics raised in the sketch, in the interactive session, and in the printed materials. The average ratings by sketch of the first two, on a five-point scale, is above 4 (see table 1 for these results). The average rating of the printed material is somewhat lower (about 3.6). None of these six ratings reveals a gender difference in audience members' ratings.

In contrast, the next three questions ask about the degree to which the issues raised reflected audience members' personal experiences, experiences of "my colleagues," and "behaviors/issues I have observed at UM." Ratings of these items average 2.80–3.53 for men, and 3.38–3.91 for women; all of these gender differences are highly significant statistically.

Finally, the last two items ("The audience/actor interactive discussion enhanced my understanding of the issues" and "The balance between giving information and encouraging discussion in the presentation as appropriate") yielded high ratings (averaging 4.0 or higher) for both men and women, with only one significant gender difference.

Before the tenure sketch (called "The Fence") was rolled out, we revised the items in our evaluation questionnaire, and decided to collect more qualitative data. The five closed-ended questions are variants of the previous ones and are also included in table 1. We only have data from ten women and seventeen men (for a 36% response rate), but the ratings are uniformly high (all above 4.15, and the overall effectiveness of the sketch 4.85). There were no gender differences in response, perhaps because the ratings were so high, but also perhaps because of the changes in the item wording.

Overall, then, the quantitative data suggest that the sketches are valued highly by both male and female audience members, but female audience members find two of them more personally resonant. Though useful for identifying overall responses, these data are not helpful in determining what long-term effects the sketch might have on those who have seen it. Some of the most revealing data we have along these lines were collected by simply asking key informants to respond via nonanonymous email queries about what worked best, what worked least well, and how to make performances more useful in the future. Some examples of responses to email about the faculty meeting sketch will offer some flavor of the responses. One male faculty member wrote,

> What I found myself thinking about most after the skit was the issue of interruptions. I tend to interrupt people a lot—though it's usually to finish their sentences, not to contradict them, and I think I'm an equal opportunity interrupter, interrupting men and women equally. My reason for thinking more about this point [was that] I began to reflect on ways to make the picture "women get interrupted more" more precise. For example, how does status enter the picture? That is, are women interrupted more because they are (at least subconsciously) perceived as having lower status than men even when they have the same academic rank? Are female professors interrupted more by their grad students and postdocs (a situation where the rank differences is big enough to presumably outweigh subconscious biases) than male professors are? . . . I see I'm describing a research project . . . so I'll stop here.

Another male faculty member wrote,

> Our faculty meetings are not like that because none of our female professors can stand to come! I think that the skit raised a number of points about departmental dynamics. Certainly every member of departmental executive committees should see it. It simply helps people be aware of the pitfalls common to interpersonal communication.

A faculty member from a different department raised an interesting issue about the limitations of the cross-unit strategy:

> This play made me immediately reflect on the dynamics among faculty in my own department and of course "my" specific role in all of it. I thought a lot about this play after the evening gathering

TABLE 1. Average Ratings of ADVANCE Sketches

	Female		Male		Overall		Significance[a]
	N	Mean	N	Mean	N	Mean	
Faculty Advising Faculty							
Scale: Not useful (1) … Highly Useful (5)							
The issue/topics raised in the actors' performance of the sketch	67	4.33	25	4.08	92	4.26	ns
The issue/topics raised in the audience/actor interactive discussion of the sketch	66	4.39	25	4.16	91	4.33	ns
The printed materials provided as resources for this presentation	29	3.55	16	3.69	45	3.60	ns
Scale: Strongly Disagree (1) … Strongly Agree (5)							
The issues raised in the performance reflected my personal experiences	67	3.60	25	2.80	92	3.38	**
The issues raised in the performance reflected experiences of my colleagues	65	4.09	24	3.42	89	3.91	**
The issues raised in the performance reflected behaviors/issues I have observed at UM	65	3.78	25	3.24	90	3.63	*
The audience/actor interactive discussion enhanced my understanding of the issues	66	4.09	25	4.04	91	4.08	ns
The balance between giving information and encouraging discussion in the presentation was appropriate	66	4.23	24	4.42	90	4.28	ns

	Female		Male		Overall		Significance[a]
	N	Mean	N	Mean	N	Mean	
Faculty Meeting							
Scale: Not useful (1) … Highly Useful (5)							
The issue/topics raised in the actors' performance of the sketch	52	4.31	117	4.05	169	4.13	ns
The issue/topics raised in the audience/actor interactive discussion of the sketch	52	4.33	118	4.09	170	4.16	ns

	Female		Male		Overall		Significance[a]
	N	Mean	N	Mean	N	Mean	
The printed materials provided as resources for this presentation	32	3.81	58	3.53	90	3.63	ns
Scale: Strongly Disagree (1) … Strongly Agree (5)							
The issues raised in the performance reflected my personal experiences	51	3.92	118	3.28	169	3.47	***
The issues raised in the performance reflected experiences of my colleagues	48	4.06	110	3.52	158	3.68	***
The issues raised in the performance reflected behaviors/issues I have observed at UM	49	3.98	114	3.58	163	3.70	*
The audience/actor interactive discussion enhanced my understanding of the issues	50	4.12	117	3.90	167	3.96	ns
The balance between giving information and encouraging discussion in the presentation was appropriate	50	4.54	117	4.20	167	4.30	*

Tenure: The Fence

	Female		Male		Overall		Significance[a]
	N	Mean	N	Mean	N	Mean	
Scale: Strongly Disagree (1) … Strongly Agree (5)							
Please rate the overall effectiveness of the CRLT sketch and interactive presentation	10	4.90	17	4.82	27	4.85	ns
The issues raised in the performance are important	10	4.90	17	4.53	27	4.67	ns
The performance made me think about some familiar interactions and situations in new ways	10	4.20	17	4.29	27	4.26	ns
The issues raised in the performance reflected issues I have observed at UM	10	4.20	17	4.41	27	4.33	ns
The audience/actor interactive discussion enhanced my understanding of the issues	10	4.50	17	4.00	27	4.19	ns

[a]ns = not significant.   *$p = .05$,   **$p = .01$,   ***$p = .001$.

but I really didn't have an opportunity to talk about it with anyone who was there for many days. I think it might be more useful to have this play performed within a department where colleagues have more opportunities to reflect informally.

In an interesting confirmation of this point, one female faculty member
wrote,

> I think this presentation is excellent, right to the point, and I find
> it way more effective than any statistics/graphs that I have seen on
> work climate for women/minorities. I can't help noticing that
> among some colleagues I spoke with, male colleagues do not per-
> ceive it in the same way as females. I have come across responses
> from shrugging shoulders to "it's a bit heavy handed, isn't it?' to
> "it was good, but our department is not like that" (not joking).
> Why that is, is probably part of the issue.

Her message was inadvertently directed to the entire group of people
who had been queried, and one of her male colleagues responded,

> I should probably confess that I am likely one of the people who
> said . . . that I found the sketch a bit heavy-handed. . . . I expect
> that the sketch was probably more powerful if you yourself have
> suffered from some (or all) of the injustices portrayed and I
> definitely should have been more sensitive to that.

He concluded his lengthy reflections by wondering about the impact of
his own behavior interrupting female and junior male colleagues:

> I guess it also made me wonder if there is any disparity in my
> behavior or if my interrupting may have a more negative impact
> on female colleagues given the general climate issues.

While it is certainly valuable to collect anonymous, aggregate data,
direct email queries have produced many responses that provide us a
richer understanding of the process of reflection during and following
the presentations.

We have learned that we cannot expect the sketches to have uniform
impact, either from one individual to the next or from one department
or college to the next. (For example, the sketches have been utilized

more often and responded to more positively in the College of Literature, Science, and the Arts than in other colleges.) Aggregate assessments are best equipped to reveal uniform outcomes, but we are also interested in finding ways to document outcomes that are unusual but important, such as the success of the faculty meeting sketch when it was performed for a single department (rather than a cross-departmental group) in LSA.

One of the difficulties involved in measuring the impact of something like the CRLT sketches is that what is easiest to measure is impact upon individuals, but what we really want to know is what impact the sketches may have had upon the entire system that is academic science and engineering at the University of Michigan. And, as with any ADVANCE intervention, it is difficult to single out effects from a single intervention when so many other interventions are taking place concurrently under ADVANCE auspices.

Conclusions

LSA's successful use of the theater sketches points to the importance of embedding such interventions in a larger agenda and engaging highly placed administrators like deans if the interventions are to have any lasting impact. Framing—giving the target audience a reason to care about and a way to make use of the information given—is crucial, as are setting and audience composition. Relatively homogeneous groups may often be best equipped to have the most constructive discussions. Those who are in a position to make tenure decisions, for example, will have a different perception of the tenure evaluation sketch than untenured faculty, who might find it threatening or overwhelming. It is important that the context in which the sketches are shown be a safe one for the audience, one that allows for receptivity and open-mindedness rather than defensiveness. Thus, despite the success of the faculty meeting sketch within one department in LSA, we still recommend showing that sketch to groups that cross departmental lines rather than using it within individual departments.

We remain open to experiment, however. And we believe that we have only begun to tap into the possible uses of interactive theater for addressing issues of academic climate. In summer 2005, CRLT held its first Summer Institute, a three-day seminar at which the players demonstrated the basics of source work, role-play, actor-audience interaction, facilitation, and other aspects of sketch creation and performance to thirty-three avid participants from sixteen other colleges and universities. The Summer Institute received rave reviews, and will be repeated. In addition, ADVANCE hopes to hold summer seminars specifically for scientists and engineers that will bring the CRLT Players and STRIDE together to mobilize faculty activists. We are certain that the CRLT Theatre Program will continue to collaborate with UM ADVANCE in finding new ways to foster discussion, reflection, and transformation in the academy.

NOTE

The authors wish to thank Constance Cook and Matthew Kaplan, director and associate director of the Center for Research on Learning and Teaching (CRLT) for their contributions, as well as Devon Dupay, assistant director of the CRLT Theater Program, and the actors who have performed in the ADVANCE sketches: Ward Beauchamp, Chad Hershock, James Ingagiola, Valerie Johnson, Omry Maoz, Melissa Peet, Hugo Shih, and John Sloan. Diana Kardia was the original facilitator and played a key role in early rehearsals and sketch creation. Thanks are also due to Steve Peterson and Chris O'Neal for their contributions. We are also grateful to Beth McGee, from Case Western Reserve University, for sharing her detailed notes from the CRLT Summer Institute with us, to Mel Hochster, Martha Pollack, Cynthia Hudgins, and Janet Malley for helpful feedback on an earlier draft, and to Keith Rainwater for evaluation data.

REFERENCES

Boal, Augusto. 1997. The theatre of the oppressed. *Unesco Courier* 50 (11): 32–36.

Bogart, Anne. 2001. *A director prepares: Seven essays on art and theatre.* New York: Routledge.

Brown, Kate H., and Diane Gillespie. 1999. Responding to moral distress in the university: Augusto Boal's theater of the oppressed. *Change,* September–October, 34–39.

Chesler, Naomi, and Mark Chesler. 2005. Theater as a community-building strategy for women in engineering: theory and practice. *Journal of Women and Minorities in Science and Engineering* 11 (1): 83–96.

Steiger, Jeffrey. 2004. *ADVANCE faculty meeting.* University of Michigan.

PART FOUR

Learning from Change

FOURTEEN

Creating a Productive and Inclusive
Academic Work Environment

C. Greer Jordan & Diana Bilimoria

DURING THE PAST two decades, women have entered science and engineering programs in unprecedented numbers. Yet they are proportionally underrepresented in tenure-track, tenured, and full professor positions within academic science and engineering (S&E) departments (Shepherd 1993; Long 2001; Valian 1999; National Science Foundation Division of Science Resources Statistics 2004). Prior research suggests that the work environment or climate of a S&E department is a key facilitator or deterrent to the advancement of women's careers (Rosser 1999; Etzkowitz, Kemelgor, and Uzzi 2000). Many studies and articles provide accounts of chilly, isolating, and hostile climates for women in academic departments (Chilly Collective 1995; Anonymous and Anonymous 1999; Rosser 1999; Etzkowitz, Kemelgor, and Uzzi 2000; Zuckerman, Cole, and Bruer 1991; Elg and Jonnergard 2003; Lawler 1999, 2002; Sonnert and Holton 1995; Stewart, Stubbs, and Malley 2002). Only the rare study addresses enabling climates and cultures for female academics. Rosser's (1999) study, for example, found that women in more cooperative and collegial science departments felt more engaged in their work, more connected to their peers, and better able to develop their professional potential.

Meyerson and Fletcher point out that "gender inequity is rooted in our cultural patterns and therefore in our organizational systems. . . . existing systems can be reinvented by altering the raw materials of organizing—concrete everyday practices in which biases are expressed"

(2003, 234–35). We found little empirical research exploring the "concrete everyday practices" of positive workplace cultures, within which both women and men are engaged and effective. Thus, we designed a case study to explore the interactions that appeared to produce a positive, supportive work culture for women and men.

In this chapter, we will introduce the study[1] we conducted and draw from our findings to present suggestions for intervening in academic departments. We believe faculty and chairs in particular will find these ideas useful in influencing change toward a more inclusive and productive workplace culture within their departments.

The Case Study

The setting of our study was an academic science department within a private Midwest university. This department had achieved a reputation for cooperation, inclusion, research productivity, and successful teaching and training of students and postdocs. In addition, over its fifteen-year history, this department attracted, retained, and advanced women throughout the faculty, including two female chairs, one an Academy of Sciences inductee.

Edgar Schein defines climate as "the feeling that is conveyed in a group by the physical layout and the way in which members of the organization interact with each other, with customers, or with other outsiders" (1992, 9). Work climates have been linked to important organizational outcomes such as satisfaction, productivity and performance, retention, and emotional support (Carr et al. 2003; August and Waltman 2004). Climate and the practices, activities, and processes of a social system constitute its culture (Schein 1992). We wanted to understand the ways in which members interacted with each other that resulted in a department culture that female and male members described as "cooperative," "collegial," or "a good place to do science." To accomplish this, we designed our study using an exploratory case study approach (Yin 2003). Our guiding question was, "How is a cooperative, inclusive, productive work culture created and embedded?" To answer this question, we employed multiple qualitative methods. Archival data were used to explore the department's history, membership, and outcomes. We observed (Bickman and Rog 1997) seminars, faculty meetings, routine lab work, and casual interactions in halls and offices, and recorded descriptive notes. Semistructured interviews (Knight 2002) were conducted with twenty-nine department members. Participants were asked to talk about (1) what brought them to work in or with the department, (2) what they found most engaging about their work and how others in the department helped them pursue that work and their overall career, and (3) who becomes a successful academic scientist and how.

We interviewed all sixteen primary faculty members, three of whom were women. Among these women, there were two full professors and one associate. There were four male full professors, seven male associate professors, and two male assistant professors. We interviewed four secondary faculty members. This sample consisted of two women at the associate rank and two men, one at the associate rank and the other at assistant rank. We interviewed three administrative and research staff, all female. We also interviewed six doctoral students and postdocs. These six participants mirrored the overall evenly divided sex representation within the student and postdoc population. They also represented a mix of large and small labs, and the new student, advanced student, and postdoc stages.

Three participants identified themselves as minorities, one black, and two Hispanic, one of whom was a faculty member. Three faculty members, all male, were from Western Europe. Half of the students and postdoc participants were U.S. born, the remaining three, all males, were from Asia, Western Europe, and South America respectively.

We recruited faculty via email and telephone for interviews. We recruited staff, students, and postdocs via email, direct mail, and personal invitations in private locations, since most did not have their own phones on campus.

We were exploring how a phenomenon presented itself through interactions; thus, we used the grounded theory approach to data collection and analysis of data (Glaser and Strauss 1967; Wells 1995). Through our collection and analysis of the interview, observation, and archival data, we identified values and beliefs, day-to-day interactions, activities, practices, and processes that appeared to contribute to the creation of a departmental culture that supported the development and advancement of female and male scientists. We have organized our findings in the form of recommendations or suggestions for faculty members and departmental leaders who would like to create a productive and inclusive academic work environment in their departments.

Suggestions for Faculty Members

Check Values and Beliefs about Who a Scientist Is and What a Scientist Does

Values and beliefs about human differences influence the extent to which an organization is able to integrate difference and increase performance (Ely and Thomas 2001). Awareness of one's own value and belief orientations is important to avoid gender schemas that can unconsciously affect views of merit and gender in science (Valian 1999). Within the department we studied, faculty professed and acted consistently with certain values and belief orientations that supported more gender-integrated conceptions of who a scientist is and what a scientist does. As illustrated below, faculty members' values and beliefs about good science influenced their subsequent actions, particularly the level of engagement in constructive interactions.

Department faculty valued "good science," which they generally defined as the pursuit of answers to questions that produced new knowledge for the field in substantive, significant ways. In addition, several faculty members, including well-established male and female scientists, saw developing future scientists as part of good science. "Good science" did not involve tweaking the work of others, chasing "popular" topics, rushing to be first to publish results at the expense of accuracy, or "scooping" ideas from others (discussed by male and female faculty members, all tenured).

Faculty also believed that both men and women could do high-quality science if they could learn quickly and were well trained, excited about science, and willing to work hard. Neither graduate students nor postdocs reported feeling stereotyped into certain roles or lines of scientific inquiry. Male and female students and postdocs reported similar experiences of faculty support of their work and life circumstances. Several faculty members cited the importance of providing developmental support to others and acted on this belief. For example, a female postdoc reported that her principal investigator, a male full professor, invited her to attend a conference, and supported her in writing her first grant for funded research. Department faculty also explicitly assessed their own ability and willingness to support the development of new junior faculty before extending job offers.

Explore Ways to Make the Department a Scientific Community

Department scientists spoke about ideas of community, concern for the department as a whole, and the need to draw all scientists into relationships with other scientists. Their belief that a good department makes for good science motivated them to lead, encourage participation of other scientists, and support activities that result in constructive interactions. One female associate professor noted, "I think he or she [a good scientist] has to be an interactive person to make the group better. You know they can't just sit in their labs and be great scientists and never talk to other people. It is good scientists that participate in group activities that have a broader impact on the department and university, because they transmit their ideas to students, postdocs, and other faculty members in the department." A male associate professor explained, "The thing that makes the department different from being sixteen independent entities is that there's interaction and there can be guidance. There can be support between these self-contained laboratories. . . . I think the better the department is, the more cases there are of faculty working together on things that benefit the department but not necessarily an individual faculty member exclusively."

Viewing the department as a community enabled department members to identify what they could learn and share with each other to advance the work and success of the department as a whole. The more resources are shared, passed along, or made widely available, the more resource-rich and attractive the work and social environment of the department. The "ladder story" perhaps best encapsulated the vision of community within the department we studied. Two faculty members, one male and one female, recounted the story that a senior male professor told them. The male associate professor's version was this: "When I came here, when I interviewed here, a professor told me a story of the department's ladder. It turns out that three or four of the faculty got together and bought an extension ladder for cleaning their gutters. And every fall they'd drive it around to their different homes and help each other do their gutters." The faculty member understood from this story that while professors have separate labs, in this department they gather and share resources that support the success of everyone's lab and the department as a whole.

Strive to Develop and Engage in Behaviors That Support Constructive Interactions

We found a set of interaction behaviors among the faculty that appeared to support inclusive processes and thus productive work outcomes. We termed them *constructive interactions*. Women and men reported these interactions across rank and nationality, and we observed other interactions in the course of day-to-day department activities. Constructive interactions consist of the following:

1. *Collegial interactions:* mutually respectful, civil, regular exchanges that occurred between department members in formal and informal settings. Collegial interactions do not require much time, or deep personal interest or knowledge of another person. However, several participants reported that collegial interactions during preemployment visits gave them the impression of a positive work climate and set expectations for further positive interactions. Hence, collegial interactions provide the basis for more complex interactions. Examples of these interactions are greetings in the hallway, small talk before meetings, or listening attentively while others speak. Departmental settings that provided a context for interactions included time before and after department presentations, faculty meetings, or the hall between offices.

2. *Tacit learning interactions:* observation opportunities and knowledge-sharing exchanges that allowed faculty at all levels to learn about the work and convey and maintain the expected work norms and behaviors. These interactions occur in the context of shared tasks and require more time in contact. Intentionally sharing learning and experiences also requires an interest in developing others. Examples of these interactions included sharing thought processes around a research question or technique, or working directly with peers, students, or postdocs to show them how work is done. Department-wide activities that supported these interactions were team teaching, committee work, and department presentations.

3. *Relational interactions:* interactions that helped people learn about and support each other on a personal and professional level. These interactions move beyond collegial interactions to professional and personal friendships. Relational interactions build trust and concern for others that can lead to significant expenditure of time in exchanging resources (Bouty 2000). Examples of these interactions included consoling others about a lab mishap, offering support and advice after loss of a grant, or helping with maintenance tasks (see "ladder story" later in this text). The activities that provided a context for these interactions included open informal gatherings of faculty, the department retreat, department picnic, "beer hour," and informal group dinners and with peers and potential new faculty.

4. *Generative interactions:* problem-solving interactions involving the sharing of resources or working to create additional resources for the benefit of the group. Department members tended to either pass on

the fruits of these collaborations to others, make the effort to give back to the group in some way, or make a new resource available to everyone. Examples of these interactions included offering techniques to colleagues, generous sharing of materials and expertise, obtaining funding for equipment for the department use, and team support of the writing of funding proposals for colleagues. The activities that provided contexts for these interactions included dyadic, small group or cross-lab meetings, department presentations and informal activities such as beer hour.

Two kinds of behavior appeared to play a prominent role in supporting constructive interactions. The first is civility, defined as "well-mannered behavior toward others" and "a courteous act or courteous acts that contribute to smoothness and ease in dealings and social relationship amenity."[2] The collegial interactions we observed consisted of civil behaviors that ranged from friendly greetings in passing to showing openness, attention, and interest while others spoke. One faculty member speculated that respect was important to the civil behaviors observed in a meeting: "You have to have the respect for each other. When you get that, then you listen to what other people say in the meeting. . . . You may not agree with them because you realize they're looking at something in a different way than you would look at it, but you can't just say, 'Well, that doesn't count.' Or 'That's not important'" (female full professor).

Second, citizenship behaviors support the constructive interactions, activities, and processes that constitute an inclusive work culture. Bolino and Turnley define citizenship behaviors as including "a variety of employee behaviors, such as taking on additional assignments, voluntarily assisting other people at work, keeping up with developments in one's field or profession, following company rules even when no one is looking, promoting and protecting the organization, and keeping a positive attitude and tolerating inconveniences at work" (2003, 60). Department members reported a variety of experiences of citizenship behaviors in the course of engaging in constructive interactions. For example, a female full professor noted, "You walk down the hall and you talk to somebody about something they're doing and they tell you. And that gives you ideas of things to do on your own, or [knowledge of] new techniques. People have been helpful about sharing new techniques, ideas, knowledge, and their technical expertise. That's really important."

Support Opportunities to Engage in Constructive Interactions across Demographic or Functional Differences

Most members of this department appeared to be particularly adept at engaging in and sustaining constructive interactions within their own labs and across the department. To promote inclusion, faculty need to actively seek to extend participation in constructive interactions to women, minorities, junior faculty, and even those with different research approaches or topics. Etzkowitz, Kemelgor, and Uzzi (2000) point out that men in academic departments have long maintained networks of interactions that support and advance their work. Unfortunately, women have often found themselves outside of the networks in which these interactions occur. It is important for women, minorities, and new faculty members to make the effort to interact with colleagues (Flynn, Chatman, and Spataro 2001). However, males or more senior faculty must also be willing to extend invitations to interact and respond to others' efforts to connect. A senior male professor reported he regularly extends an invitation to all faculty members to gather, relax, have a beer, and talk. He personally extended an invitation to these gatherings to a new female faculty member to ensure that she knew she was welcome. A junior male faculty member reported how important relational interactions with a more senior faculty member were during his first year: "I think another person who was really important for me at the beginning was [a male associate professor] because he made me feel really welcome at the beginning. So he took me around a lot, and he encouraged me after my grant was rejected. He said, 'You don't have to worry about it. That's the normal way it goes . . .' I mean, at least for me, I needed some positive reinforcement there, to make it through." A female associate professor recounted engaging in hallway conversations about science with her senior peers that helped excite and motivate her early in her career.

Resist Pressures of Career Unimodality

A senior male professor noted that scientists in the department tended not to be "unimodal"—intently focused on only one aspect of science for career satisfaction and identity. He noted that scientists in the department exercised their love of science in a variety of ways including research, teaching, community outreach, training and development, or administration. Some scientists, both male and female, made adjustments to work roles and schedules to spend time with their children. Two senior professors willingly spent time counseling students during their lab placements or coordinating graduate course content.

Both men and women reported institutional pressures toward a single focus on obtaining funding, which conflicted with high-quality science and the responsibility they felt for developing others, whether in the lab or at home. Despite the tensions, a multimodal approach to a scientific career better supports inclusion, engagement, and productive outcomes for many department members. Self-verification theorists suggest that people actively strive to ensure that their experiences in groups confirm their self-views. Thus, an individual is attracted to a group as far as members of that group confirm her or his identity. Successful cooperation in diverse groups depends on how well group members can affirm each other's identities (Polzer, Milton, and Swann 2002). The implications of these findings are, first, that an individual who has a narrow conception of her or his own identity will have more limited opportunities for self-identity confirmation and less ability to confirm the identities of others who are different. The wider a person's interests and picture of who she or he is, the better the chance of finding satisfaction in some aspect of academic work and the more options for connecting with others. Second, the wider the range of identities that can be confirmed within a group, the greater a group's capacity for including and leveraging difference. Multimodality within a group can support different identities and associated interests and thus, inclusion and engagement of diverse constituencies.

Suggestions for Departmental Leadership (Chairs)

Leadership is theorized to be a key factor in the development and maintenance of an organization's culture (Schein 1992). Across the board, several leadership practices, employed by both chairs, enabled constructive interactions in the department. These practices were also recognizable, to varying degrees, among faculty members in their individual labs.

Relate to Faculty Fairly and Equitably

Several faculty specifically cited the fairness and forthrightness of the current chair. One male associate faculty member said, "I liked the department chair. She had the reputation of being an absolute square dealer and nothing that I encountered during the process of recruitment, or since, has changed my mind. And I think that an important criterion for what makes a department work is that people are able to trust what they hear, especially from the Chair. And they have to have a sense that the Chair is fair. I think [the Chair] is exceptionally good at both of those things. I've never felt that anything she said wasn't something that I couldn't bank on." Thus, faculty could trust that they were receiving all the available information about a situation. They could also trust that the chair's interests extended beyond her own lab to the advancement of everyone's labs. One male associate professor noted, "I think that the chair tries to make [demographic or personal characteristics] as neutral as possible. The chair is about supporting you to produce as good a science as you can in your own lab. You can come into work with a Mohawk or do whatever you want to do. I don't think that she'll devote one iota of her energy to worrying about that, compared to if you insist on pursuing something really useless and waste a year chasing something that turns out to be nonsense."

Create and Maintain Open Information and Transparent Decision-Making Processes

Pelled, Ledford, and Mohrman (1999) identified decision-making influence and access to sensitive information as two key indicators of workplace inclusion. Interaction behaviors of the chair made open communication processes, such as faculty meetings, effective in creating inclusion. The researcher observing a faculty meeting noticed that the chair started discussion by providing data, including relevant information from the external environment, and then made clear focused requests for input. After this, she spent a significant amount of time listening. She also asked for input from faculty who were silent but possibly affected by the decision. The chair maintained an open, respectful stance toward faculty comments throughout the meeting. Faculty supported this process as well, by asking clarifying questions instead of attacking one another, separating data from opinion, making timely observations about the group's process and use of nonpersonal, tension-dissipating humor.

Depending on the proximity of faculty members, communication culture and the leadership style of the chair, face-to-face faculty meetings may not be a good fit. For example, the previous chair, operating in a smaller department, approached faculty one-on-one. Alternatively, for some groups email or conference calling might be effective means of facilitating information flow. Regardless of means, the objective must be to find ways and means of communication and feedback that engage everyone in information-sharing and decision-making processes.

For example, in the department we studied, all primary and active secondary faculty members, not a committee, participated in the faculty recruiting and selection process. The process included group assessment of candidate credentials before interviewing and rotating the faculty who took candidates to meals, escorted them between interviews, or took them on tours. Faculty also attended the candidate "chalk talks" and presentations. Everyone's observations from these activities were combined and discussed in a meeting of the faculty. This process ensured transparency in recruitment and selection. There was no mystery around who was involved in the selection of a new faculty member or how a newcomer could fit into the department. A single individual or subgroup (e.g., senior professors, professors of certain standing in the field, or an age or gender subgroup) did not monopolize decision-making power. In the words of one female full professor, "I think that everybody knew their input counted. In the end, we did go the way that the group decided for all the positions." Transparency also facilitated trust and engagement between department members. One male full professor commented, "So there aren't any politics, and nobody's being forced to do things. People are genuinely interested in teaching or are certainly interested in the job search. And so it's sort of a team effort, which makes it rewarding. I think that there is not very much of a hierarchy in the Department, between the junior faculty and the senior faculty. And, to some extent, the students feel like they're part of the process. So people feel empowered. People's opinions are asked and they receive feedback."

The team recruiting process also facilitated integration of the new member into the department. A female full professor commented, "Recruiting as a group is important because you want to bring in people that everybody feels good about." Feeling good about a person promoted interest in that person's success and encouraged further acts of inclusion and development. A male associate professor noted, "The strength of the department is that it's got a large group of faculty that has been involved in hiring the people. [These faculty] are now invested in many people in the department because they played key roles in their recruitment. And so we're trying to work on ways, through the infrastructure of the department, to expand the circle."

Use the Role of Chair in Service to the Department

How the chair handled her role appeared to influence the level of trust and openness to cooperation in the department. Both chairs in the department we studied appeared to be adept at separating their role as chair from the advancement of their own labs. Neither appeared to have status or self-promotion as an objective, but instead viewed the role of chair as service to the department. Thus, they could disseminate information and involve the entire faculty body in discussion of important issues without feeling personally threatened. Their openness also prevented fear and competition over information and access to important resources.

There were several indicators of this service stance. Participants reported instances of the chairs assisting or even securing funding for new faculty. Neither chair treated the department as an extension of her self or her own work by monopolizing resources and recognition for their own ends. They did not use their status to demand unwarranted resources, authorship, or access. As an example, a junior faculty member noted that the chair did not question his primary authorship of work that she had supported, which was contrary to the experiences of other junior faculty he knew at other schools.

Thus, both chairs appeared to view the role of chair as service to the department and the advancement of a scientific community, not as a reward to leverage for their own benefit.

Recruit For and Encourage Faculty Development That Supports Constructive Interactions

Recruiting is a mechanism that is "one of the most subtle yet most potent ways through which cultural assumptions get embedded and perpetuated" (Schein 1992, 243). Bolino and Turnley suggest that the way to foster more civil, collegial, and cooperative culture is to select for employees who practice citizenship behaviors. Bolino and Turnley explain: "Employees who are conscientious, optimistic, extroverted, empathetic, and team-oriented may be more willing to engage in certain types of citizenship behaviors" (Bolino and Turnley 2003, 60). They go on to suggest that evaluation of new employees include assessment of proclivities in these areas. Chairs and faculty need to not only look at the academic, research, and teaching credentials of candidates but also their interpersonal skills, team orientation, and their values and beliefs about science and who can do science.

The initial group of six scientists who formed the department we studied valued cooperation, collaboration, and a strife-free work environment. A secondary faculty member observed that the department appeared to consist of good scientists who wanted to "work in peace." Faculty invited new people to join the department who would most likely value and engage in constructive interactions. Recruiting appeared to be essential to sustaining an environment in which constructive interactions could take place.

Bolino and Turnley also suggest training programs and faculty development opportunities that teach cooperation and build relationships (Bolino and Turnley 2003, 65). Such training can shed light on unconscious habits or behaviors that may not encourage constructive interactions. Coaching and training works best when participants have a compelling goal, emotional support from peers, and opportunities for safe experimentation and feedback (Roberts et al. 2005; Druskat and Wolff 2001). The best place to start with training may be with the formal and informal leaders of the department who can play a key role in supporting long-term change.

Create and Support a Variety of Status-Leveling Contexts for Constructive Interactions within a Department

Various voluntary activities, which bring people together at different times and in different contexts, provide opportunities for interaction. Constructive interactions need supportive contexts. Variety in timing and contexts also allows people to fit events into their schedules and choose those with which they are most comfortable. Assuming that department members have respect for and interest in one another and behavioral skills in interacting, the following activities provide opportunities for constructive interactions between faculty, students, postdocs, and staff.

Team Teaching across Faculty Ranks. Faculty shared responsibility for the design and delivery of graduate courses. Various faculty members, across ranks, participated in teaching parts of the graduate program. A senior faculty member coordinated this activity. He explained, "I give some of the lectures in the course [graduate level science course], but I also organize everything like the exams and the handouts and grading, etc. Quite a few people in the Department cooperate. About six different people give lectures that have to be coordinated. It's a very positive experience. People are very willing to do it and they meet deadlines that

I set for them and do their best. And the students seem to like the course." Advantages of team teaching mentioned by faculty included a manageable teaching load, particularly beneficial for junior faculty trying to establish a lab, opportunities for junior faculty to learn from experienced faculty, and opportunities to interact with different faculty.

Department-Level Social Events. Both chairs of the department we studied initiated or supported creation of department-wide social activities that afforded faculty, students, and postdocs opportunities to interact outside of their labs. Faculty members have supported these initiatives by participating in and rotating the leadership of activities. Activities include a weekly beer hour, which a faculty member described as a "sort of science happy hour." The current chair also introduced an informal department picnic and retreat.

Regular Meaningful Seminars and Presentations. Many faculty members mentioned the importance of department seminars and presentations in stimulating ideas, helping them to fashion their own projects, and making contact with peers with mutual interests. Two female students indicated that the interactive, interesting, and well-attended research seminars attracted them to join the department. The faculty emphasized the importance of these seminars for the development of young scientists by making the sessions mandatory for graduate students. Faculty members, both primary and secondary, were well represented at the session we observed. The room was abuzz with conversation even before the presentation. Faculty members were responsive to the presenter, asking questions that helped the speaker clarify points or consider new angles or ideas about the research. Afterward, some faculty lingered, talking with peers and students. Thus, seminars and presentations are an important context for constructive interactions.

Team teaching, department social events, and high-quality research seminars and presentations provide department members with the knowledge and information they need to advance their work. Participants reported access to role models for approaches to the work, peers they could generate ideas with, and important new techniques and methods being available for the asking. These activities support dissemination of information, tools, and materials—all strategic, work-advancing resources. One male associate professor said, "If you had questions, you could go talk to one another very freely. You could ask people for advice, people that were more senior to me. I found it very harmonious and productive in a cooperative environment."

Conclusions

The suggestions presented in this chapter provide faculty and chairs with starting points for making changes that can influence the environment of a department. Many will read this chapter because they are interested in opportunities to improve a department's work environment or culture. Perhaps they have observed a department's failure to recruit more women, or to advance them to tenure. Perhaps they have found that women's pay is not on par with men's of comparable accomplishment, or that women have been leaving the field in disproportionate numbers. These are traditional indicators of a "gender inequity problem" as described by Meyerson and Fletcher (2003, 235). Creating an inclusive, productive department culture requires a coordinated, cooperative effort, utilizing many of the guidelines suggested in this chapter. Meyerson and Fletcher (2003) recommend a "small wins" approach. The small-wins strategy "creates changes through diagnosis, dialogue, and experimentation. It benefits not just women, but also men and the organization as a whole" (2003, 236). The first step is diagnosis or data gathering about the current situation. Meyerson and Fletcher briefly explain the entire process as follows: "People must get together to talk about the work culture and determine which everyday practices are undermining effectiveness. Next, experimentation begins. Managers can launch a small initiative—or several at one time—to try to eradicate the practices that produce inequity and replace them with practices that work better for everyone. Often the experiment works—and more quickly than people would suspect. Sometimes it fixes only the symptom and loses its link to the underlying cause. When that happens, other incremental changes must be tried before a real win occurs" (2003, 235).

Below are several questions that provide a starting point for diagnosis of a department's current culture as part of a small-wins effort:

1. What gender schemas appear to be operating about who can do science, and are there any associated double standards operating in the department (Valian 1999)?

2. Are there misunderstandings and relationship tensions that might be related to different identities and expectations around interacting (Flynn 2005)?

3. Are there uncivil behaviors that block constructive interactions (Pearson and Porath 2005)?

4. Are there possible synergies between research approaches, methods, techniques, equipment, or other needs that can highlight interdependencies and provide incentives for interactions?

5. Who has access to important information? Who has decision-making influence and why (Pelled, Ledford, and Mohrman 1999)?

6. What characteristics, skills, and abilities are considered during the hiring process? The promotion and tenure process? What does this say about the values, beliefs, and culture of the department? What does this say about who is likely to join, succeed, or fail in this department?

Thinking through the answers to these questions can help raise awareness of the tacit and covert aspects of a department's culture. However, we caution that getting the answers to such questions requires conditions of trust, thoughtful and honest consideration, and openness to learning. An approach to change that can create favorable conditions for inquiry with respect to diversity is to make *learning and integration* the goal of initiatives (Thomas and Ely 2001; Ely and Thomas 2001). Learning new ways to expand and leverage the skills and competences of department members can benefit everyone. Otherwise, the move toward constructive interactions can become an exercise in political correctness or a justice-and-fairness crusade that may alienate minority and majority members of the department and yield little substantive inclusion or productive resources for the department.

NOTES

This study was supported by an NSF ADVANCE Institutional Transformation Grant, SBE-0245054, start date September 1, 2003, end date August 31, 2008.

1. See Bilimoria and Jordan 2005 and Jordan and Bilimoria 2005 for details about the study and the resulting theoretical model.

2. *Roget's II: The New Thesaurus,* 3rd ed., s.v. "civility."

REFERENCES

Anonymous and Anonymous. 1999. Tenure in a chilly climate. *Political Science and Politics* 32 (1): 91–99.

August, Louise, and Jean Waltman. 2004. Culture, climate, and contribution: Career satisfaction among female faculty. *Research in Higher Education* 45 (2):177–92.

Bickman, L., and D. Rog. 1997. *Handbook of applied social research methods*. Thousands Oaks, CA: Sage.

Bilimoria, Diana, and C. Greer Jordan. 2005. A good place to do science. http://www.case.edu/admin/aces/documents/science_department.doc.

Bolino, Mark C., and William H. Turnley. 2003. Going the extra mile: Cultivating and managing employee citizenship behavior. *Academy of Management Executive* 17 (3): 60–72.

Bouty, Isabelle. 2000. Interpersonal and interaction influences on informal resource exchanges between R&D researchers across organizational boundaries. *Academy of Management Journal* 43 (1): 50–56.

Carr, Jennifer Z., Aaron M. Schmidt, J. Kevin Ford, and Richard P. Deshon. 2003. Climate perceptions matter: A meta-analytic path analysis relating molar climate, cognitive and affective states, and individual level work outcomes. *Journal of Applied Psychology* 88 (4): 605–19.

The Chilly Collective, ed. 1995. *Breaking anonymity: The chilly climate for women faculty*. Waterloo, Ontario: Wilfrid Laurier University Press.

Druskat, Vanessa Urch, and Steven B. Wolff. 2001. Building the emotional intelligence of groups. *Harvard Business Review* 79 (3): 80–90.

Elg, Ulf, and Karin Jonnergard. 2003. Inclusion of female students in academia: The case of a Swedish university department. *Gender, Work and Organization* 10 (2): 154–74.

Ely, Robin J, and David A Thomas. 2001. Cultural diversity at work: The effects of diversity perspectives on work group processes and outcomes. *Administrative Science Quarterly* 46 (2): 229–73.

Etzkowitz, Henry, Carol Kemelgor, and Brian Uzzi. 2000. *Athena unbound: The advancement of women in science and technology*. New York: Cambridge University Press.

Flynn, Francis. 2005. Identity orientations and forms of social exchange in organizations. *Academy of Management Review* 30 (4): 737–50.

Flynn, Francis J., Jennifer A. Chatman, and Sandra E. Spataro. 2001. Getting to know you: The influence of personality on impressions and performance of demographically different people in organizations. *Administrative Science Quarterly* 46 (3): 414–42.

Glaser, Barney G., and Anselm L. Strauss. 1967. *The discovery of grounded theory: Strategies for qualitative research*. Hawthorne, NY: Aldine de Gruyter.

Jordan, C. Greer, and Diana Bilimoria. 2005. In pursuit of good science: The social process of creating and embedding an inclusive, effective science work environment. Working Paper WP-05-05, Weatherhead School of Management, Case Western Reserve University.

Knight, Peter T. 2002. *Small-scale research: Pragmatic inquiry in social science and the caring professions*. Thousand Oaks, CA: Sage.

Lawler, Andrew. 1999. Tenured women battle to make it less lonely at the top. *Science* 286:1272.

———. 2002. Women in academia: Engineers marginalized, M.I.T. Report Concludes. *Science* 295:2192.

Long, J. Scott, ed. 2001. *From scarcity to visibility: Gender differences in the careers of doctoral scientists and engineers*. Washington, DC: National Academy Press.

Meyerson, Debra E., and Joyce K. Fletcher. 2003. A modest manifesto for shattering the glass ceiling. In *Reader in gender, work, and organization,* ed. by R. J. Ely, E. G. Foldy, M. A. Scully and The Center for Gender in Organizations Simmons School of Management, 230–41. Malden, MA: Blackwell.

National Science Foundation Division of Science Resources Statistics. 2004. *Gender differences in the careers of academic scientists and engineers, N.S.F. 04–323.* Arlington, VA: National Science Foundation.

Pearson, Christine M., and Christine L. Porath. 2005. On the nature, consequence and remedies of workplace incivility: No time for "nice"? Think again. *Academy of Management Executive* 19 (1): 7–18.

Pelled, Lisa H., Gerald E. Ledford, and Susan A. Mohrman. 1999. Demographic dissimilarity and workplace inclusion. *Journal of Management Studies* 36 (7): 1013–31.

Polzer, Jeffrey T., Laurie P. Milton, and William B. Swann Jr. 2002. Capitalizing on diversity: Interpersonal congruence in small work groups. *Administrative Science Quarterly* 47 (2): 296–324.

Roberts, Laura Morgan, Jane E. Dutton, Gretchen M. Spreitzer, Emily D. Heaphy, and Robert E. Quinn. 2005. Composing the reflected best-self portrait: Building pathways for becoming extraordinary in work organizations. *Academy of Management Review* 30 (4): 712–36.

Rosser, Sue V. 1999. Different laboratory/work climates: Impacts on women in the workplace. *NY Academy of Science* 869 (1): 95–101.

Schein, Edgar H. 1992. *Organizational culture and leadership.* 2nd ed. San Francisco: Jossey-Bass.

Shepherd, Linda Jean. 1993. *Lifting the veil: The feminine face of science.* Boston: Shambhala.

Sonnert, Gerhard, and Gerald J. Holton. 1995. *Who succeeds in science? The gender dimension.* New Brunswick, NJ: Rutgers University Press.

Stewart, Abigail J., Julie R. Stubbs, and Janet E. Malley. 2002. Assessing the academic work environment for women scientists and engineers. http://sitemaker.umich.edu/advance/reports__publications__and_grant_proposals#climate

Thomas, David A., and Robin J. Ely. 2001. Making differences matter: A new paradigm for managing diversity. In *Harvard Business Review on Managing Diversity,* 33–66. Boston: Harvard Business School Press.

Valian, Virginia. 1999. *Why so slow? The advancement of women.* Cambridge: MIT Press.

Wells, Kathleen. 1995. The strategy of grounded theory: Possibilities and problems. *Social Work Research* 19 (1): 33–37.

Yin, Robert K. 2003. *Case study research: Design and methods.* 3rd ed. Thousand Oaks, CA: Sage.

Zuckerman, Harriet, Jonathan R. Cole, and John T. Bruer. 1991. *The outer circle.* New Haven: Yale University Press.

FIFTEEN

Advancing Women Science Faculty in a Small Hispanic Undergraduate Institution

Idalia Ramos & Sara Benítez

WHEN NSF announced the "first round" of ADVANCE Institutional transformation awards in 2001, we immediately realized that our institution was quite different from the other awardees (Hunter College, Georgia Institute of Technology, New Mexico State, University of Colorado, Boulder, University of Michigan, University of California, Irvine, University of Washington, University of Wisconsin, Madison). We are an undergraduate, small university with a high percentage of female science students. Since the award was made we have had the opportunity to meet with our colleagues from other undergraduate, small, and particularly minority institutions and have found that they confront many of the challenges we have at the University of Puerto Rico–Humacao (UPRH). These challenges can be summarized in the following question: How can women in science advance in an institution where advancing seems difficult for all faculty (including men)?

UPRH is part of the University of Puerto Rico (UPR) public university system. It offers nineteen bachelor degrees and has 4,300 students and 340 faculty members. Ninety-nine percent of the students are Puerto Rican, and over 70% are female. The natural science programs at UPRH were developed in the 1970s and 1980s to train workers for the industries in the Eastern Region of Puerto Rico. UPRH is among the most productive science and mathematics institutions in Puerto Rico in

243

graduating seniors. Recent classes produced more chemistry, physics, and mathematics baccalaureate graduates than any other institution on the island.

If the faculty of a university should mirror the diversity of the community it serves (CAWMSET 2000; National Science Foundation 2000), then the number of women in the faculty at UPRH is quite low compared to the number of female students or to potential women faculty as evidenced by the high percentage of female undergraduate students in science programs in Puerto Rico. The distribution of women faculty in each of the ADVANCE target programs (figures 1 and 2) illustrates this situation. In the Department of Social Sciences, 73% of the students are female, while the percentage of female faculty is 42%. In Biology the percentage shrinks from 72% to 46%, from 67% to 37% in Chemistry, from 46% to 32% in Mathematics, and from 27% to 14% in Physics. In Biology, the only department with multiple bachelor's degree programs, the distribution of women faculty in certain programs is not as great as when examining the numbers for the whole department. As shown in figure 2, women are represented on the faculty in Microbiology and General Biology in numbers equal or close to that of female students: 75% female faculty and 75% female students in Microbiology, and 72% and 69%, respectively, in General Biology. However, we observe a drastic difference in the other two programs; from 69% women students to 25% women faculty in Marine Biology, and from 60% to 0% in Wild Life Management (University of Puerto Rico–Humacao Planning Office 2004).

The working conditions for faculty at UPRH include a high teaching load of twelve credits per semester (typically four courses). Salaries for faculty in Puerto Rico tend to be lower than on the U.S. mainland and are based on a fixed scale that considers academic degree, rank, and years in the position. There is a higher scale for physicians, architects, lawyers, and engineers. To increase their income, faculty usually accept additional teaching loads for a fraction of regular time salary. Administrative position bonuses are also used to increase salary. Nevertheless, these additional functions do not normally help faculty toward promotions nor to reach leadership positions. At UPR, top administrative positions are tied to governmental politics, as the president of the UPR is elected by a board of trustees, the members of which are designated by the governor. Every time the political party in power changes, the university's high-level administrators are replaced and many academic projects are discontinued. As a result, top administrative positions are less

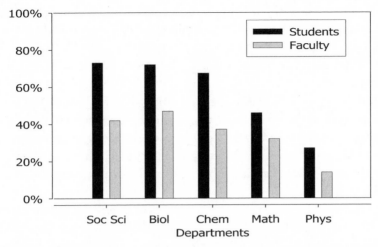

Figure 1. Distribution of Women (students and faculty) in Science Department at UPRH

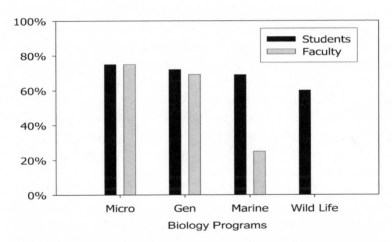

Figure 2. Distribution of Women (students and faculty) in Biology Programs at UPRH

prestigious than in other institutions because they are more political than academic.

In most science departments at UPRH, research with undergraduate student participation is recognized as an essential activity, but institutional funds for research are limited. To promote productivity, several years ago UPRH approved a salary bonus based on publications and external research funds. Faculty with higher positions and salary are those who have their own externally funded research projects. Thus an active research project can be the fastest road to a full professorship.

Faculty with master's degrees normally start as instructors; those with Ph.D.s normally start with an assistant professor rank. Faculty members qualify for promotion after a number of years in a rank (fewer for faculty with Ph.D.s). UPRH uses a defined point system for promotion. Points are awarded for each faculty activity, and criteria are published so faculty know exactly how many points are necessary for promotion for each rank. Although carrying out cutting-edge research is not required, some research activity is essential to be promoted, particularly to the higher ranks.

At UPRH faculty can move from a non-tenure-track to a tenure-track faculty position when there is an open position in the department. When a faculty member moves to a tenure-track position, the time spent in the non–tenure track "counts" toward tenure. Although this is a fair procedure for non-tenure-track faculty, the excessive number of faculty in these positions has created other problems. Faculty with these contracts do not enjoy full benefits or job security, and the work expected from them, besides teaching, is unclear. In some departments, non-tenure-track contracts have become "pre-tenure-track" contracts. This considerably limits the pool of candidates, as non-tenure-track contracts are less likely to be announced and are not attractive for highly qualified candidates. This practice has certainly affected the recruitment of women faculty in science at UPRH.

As already stated, working conditions for faculty at Humacao are challenging for all faculty members: high teaching load; lack of research funds; status as an undergraduate institution that has less equipment than do research universities; a critical lack of space for laboratories and for faculty advising; a freeze on hiring because of budget constraints; limited student research opportunities in the field of social sciences, since UPRH offers only one bachelor's degree in this area; competition for funds with larger campuses that are given higher priorities; less funding for overhead costs than is given to U.S.-based institutions, and most of the overhead is not transferred to the campus for strengthening the program. Under these circumstances, we return to the question we asked ourselves in the introduction: how can women in science advance in an institution where advancing seems difficult for all faculty? And what are the unique and basic barriers for women's advancement in our institution?

Through ADVANCE we have documented the basic and unique

barriers for women's advancement at the UPRH. Differences in academic degrees at time of hiring (fewer women hired with Ph.D.s), types
of contracts (more women in non-tenure-track positions), length of
time between promotions, in salaries based on differences in research
and publication bonuses, in space allocation and quality of the space, and
in resources between men and women—all these factors prevent
women from advancing to higher academic positions. Conflicts
between work and family, isolation, and lack of research collaborations
are also among the barriers for women faculty in science at UPRH. In
addition, excessive teaching loads and service responsibilities, as well as
lack of negotiation and time management skills, have negative effects on
women at UPRH (Valian 1999).

On the other hand, because of all the constraints faced by the faculty
at UPRH, it was important to develop a transformation and advancement program for women faculty that would clearly convey the message
that what was good for women faculty would be good for the whole
university. This situation challenged us to develop a project that would
be based on interactions among different levels and components and on
promoting collaboration.

The ADVANCE program proposed to implement an institutional
transformation process to promote the advancement of women faculty
to the highest academic positions. This would be achieved by having a
greater number of women with productive and funded research projects, space for research, and higher income and productivity through
research bonuses and research resources. ADVANCE has focused on
providing support to women faculty to increase their research productivity (funds, training, and mentoring) and on developing new policies
and procedures for faculty recruitment. Institutional research has provided relevant data on gender issues and has supported the development
and implementation of policies. To increase the future pool of women
applicants, we also mentor and provide research support to talented
female undergraduates interested in pursuing academic careers in science
(Ramos, Vega, and Valdés 2003). All these efforts have been based on
and supported with leadership development of faculty and of key players who have administrative/leadership roles, by encouraging and facilitating group interactions and by strategically organizing our work
processes in a way that integrates different university constituencies,
components, and levels to promote and sustain change in beliefs and
practices (Ramos, Hernández, and Díaz 2003).

Leadership and Collaboration: Key Players

ADVANCE proposed to increase awareness of the gender issues that inform the situation of women science faculty, who had already benefited from programs that brought an intense focus on gender issues even before the award was received. These programs helped us to develop strategies to involve the university community in the implementation of new programs and how to change the institutional climate to better support the advancement of women. In the late 1980s, a Women's Affairs Committee was organized. In 1994, a special program to recruit women in physics was established, and in 1999 a special program to reduce violence against women on campus was initiated. Before ADVANCE began in 2001, the principal investigator (PI) and co–principal investigator (co-PI) were actively involved in these efforts. Both knew many of their colleagues and administrators and were recognized as campus leaders well before their role in the ADVANCE project. The basic strategy was to build upon this awareness and to integrate different university constituencies to promote institutional change by creating an ADVANCE campus-wide Coordinating Committee that would integrate people and groups that had been working with gender issues with administrators, programs, and offices that provide services to women faculty.[1]

The Coordinating Committee was organized to obtain the collaboration of key players at the institution. It meets every semester to collaborate in the implementation and dissemination of the ADVANCE program. Members of this group also work together on a more frequent basis to further the program's objectives. The members provide leadership and support in their areas of expertise (human resources, family-work policies, library resources, academic computing, software, and so on) to the participants and to the program.

Department chairs provide the top-down leadership necessary in the implementation of the program. In the absence of formal institutional policies for faculty hiring, department chairs determine how searches are conducted, assign courses to faculty, and allocate space for research. ADVANCE has worked closely with the science department chairs so that they can identify the benefits of the program, particularly in recognizing how participating faculty elevate the prestige of their departments and how to get nonparticipating faculty to support their colleagues. They understand and have the resources to convey to the members of their departments that gender equity benefits the department as a whole.

A special committee was appointed to review procedures for faculty recruitment and to develop a plan to promote diversity. The members of the committee are leaders who have the resources to promote gender equity and diversity: the academic dean, chairpersons of the science departments, the director of the Human Resources Office, the director of the Office of Equal Opportunities, and a representative from the Women's Affairs Committee. This group has direct support from a legal advisor who is an experienced law professor and a recognized feminist. Including a legal advisor for ADVANCE was not contemplated in the original proposal, but her role has been fundamental. Although most universities employ legal advisors, they generally evaluate policies from a purely administrative perspective. Looking at policies and procedures in a context of institutional transformation for gender equity and diversity requires solid experience as faculty member, knowledge of gender issues, and expertise in diversity and human rights issues. The legal advisor has provided an educational opportunity and a training context for the institution's leaders. In the process of developing and reviewing regulations, she has not only provided legal advice but has developed a learning community that has helped produce the knowledge needed to transform the university's practices and beliefs on gender equity and diversity.

Advancement has also been promoted by an ADVANCE Advisory Board. Their role is to contribute to the assessment of ADVANCE's strategies and accomplishments and to help plan for the sustainability of the ADVANCE program beyond the funding period. The members are well-known leaders and experts from different fields in Puerto Rico and the U.S. mainland as well as knowledgeable about the culture of UPRH and Hispanic women in science.[2] They provide a highly regarded platform for interacting with faculty and administrators on gender issues. The board meets once a year in Puerto Rico during the ADVANCE annual meeting and submits a report to the PIs and the institution.

The UPRH Faculty Association has embraced ADVANCE goals as part of their commitment to improve faculty working conditions. In 2002, they approved an ADVANCE-like program on a smaller scale to provide funds for travel, publications, and other purposes for women faculty in nonscience departments. They collaborate in training and dissemination and have been instrumental in discussions about how strategies of the ADVANCE program can be implemented in other departments at UPRH.

Changes at Multiple Levels of the Organization

Mobilizing for change requires knowing the institution and its different levels and components. Moreover, it is based on strategically developed opportunities for collaboration and leadership built on clearly developed and shared goals. In a small undergraduate Hispanic university, sustained institutional changes need to improve the campus–wide climate.

Transforming the institution to better support the advancement of women requires change at multiple levels of organization, including that of the individual, the department, the upper administrative levels, and the institution as a community. Institutional transformation at UPRH targets students, faculty, nonfaculty personnel, and academic departments, as well as the institution at large. Thus, transformation results in a cultural change that can be sustained through time.

Individual. At the individual level, the program has focused on providing research support and developing skills that will allow faculty to maintain and develop externally funded research projects and reach academic leadership positions. ADVANCE research awards provide support for release time, travel, equipment, materials, research assistants, publications, visits from collaborators and mentors, and other expenses. Release time is essential for faculty to do research in a university with high teaching loads. In addition, matching funds from UPR are used to purchase small pieces of equipment.

Workshops on proposal writing and budget development have been provided on a regular basis. In addition, workshops on negotiating the promotion and tenure system, family–work conflicts, linking research and teaching, establishing research collaborations, and disseminating results have increased the faculty's knowledge of how to achieve their own goals and navigate the system.

The small research awards and training have had a strong positive impact on research productivity and have strengthened faculty's skills in obtaining additional funding. For the thirty-four women in science who have received the research awards, the support has resulted in over eighty publications. As a direct result of their research, women are able to advance through the ranks at UPRH more quickly than they would otherwise. As explained in this chapter's introduction, the faculty promotion system at UPRH is based on a point scale, and faculty who accumulate the minimum number of points for his or her rank are automatically promoted. This information is public, and faculty can choose not to apply for promotion if they know or think they will not have the required points. From personal interviews we have found that the increased research productivity due to ADVANCE support gave women faculty not only the necessary points but also the confidence to apply for promotion. As a result, in the four years after the program started (2001–5), of the fifteen promotions awarded to the science faculty, eight (53%) were to women. In comparison, during the four years prior to ADVANCE (1997–2001) of the eleven promotions awarded to the science faculty, only two were to women.

Networking to avoid isolation and for dealing with frustration has also been among the program's implementation strategies (Ginorio 1995). Through personal stories, participants have described a heightened level of self-esteem as a result of success in performing research, and of presenting their research in symposia and journal articles. A new feeling of solidarity and empowerment exists among the female faculty. Interviews with participants show that they believe that the program helped them enhance connections and collaborations with other faculty. ADVANCE strongly encouraged women faculty to establish research collaborations with other women, but also with male faculty at UPRH and other institutions. This was achieved through mentoring and was stated clearly in the grant applications. For example, in the Call for Proposals for Equipment, preference was given to proposals for collaborative projects and requests for equipment that would be shared with other researchers. Involving men as collaborators has had additional benefits. They have recognized the indirect benefits through opportunities for increased research cooperation resulting from the awards to women faculty. These benefits and collaborations have prevented backlash against the program.

Interviews with ADVANCE participants also show how women faculty from different "generations" perceive the program's effects. Recently hired women (ten years or less in 2005) believe that ADVANCE support helped them to take the lead early in their careers. Women with seniority (more than ten years in 2005) think the support from ADVANCE was fundamental, but they wish the program had come earlier in their careers. One woman stated that during the last five years she has "restarted a scientific career" that is reflected in her publications in peer review journals, but she thinks the program arrived twenty years too late in her career.

Department. Department chairs have been directly involved in many levels of the program's implementation. They are targeted leaders who are trained and are members of two important committees: the Coordinating Committee and the Regulations Committee. At the department level the ADVANCE program has brought a heightened awareness of gender issues. Workshops that have helped faculty in negotiating the point system for promotion are open to all faculty members. Another change at the department level is collaborative research and coteaching between men and women. In addition, the departmental transformation is appreciated by male faculty members because there has been change in the whole department as a result of more student researchers, multi-user equipment, and changes in recruitment processes.

This last change occurred through educating department chairs. The ADVANCE program has highlighted the need for and appropriateness of writing gender-neutral job advertisements and disseminating them widely, either through Internet posting or publication in print media that are widely available beyond the local area. In the most recent search process, this resulted in an increase in the number of qualified women candidates, from 0% to 20% in the Physics and Electronics Department and 0% to 27% in the Computer Science Program.

Institutional. The ADVANCE program has resulted in an intense focus on gender issues in part because of the small size of the institution. All faculty members are aware of the program and its accomplishments. The size of the grant awarded by NSF is substantial for a small under-funded, public, undergraduate institution. The entire UPRH community understands the benefits to all of advancing the status of women in science. Because of accumulated experience in gender issues and the small scale of the campus, institutional change has occurred quickly. Because of this new awareness, it is likely that the advancement of women in the science departments will be maintained. ADVANCE has used multiple strategies to involve the community and promote the program's goals, such as annual meetings, presentations in conferences, participation in open houses at UPRH, science fairs, and a Women in Science Photo Marathon (annual photo competition in which participants submit photos of women scientists or women working on science projects).

Support by senior administrators gives UPRH an opportunity to be a role model for other small campuses. The shift in the institution's awareness of the importance of research to faculty development, under-graduate experience, and prestige within the scientific community should eventually change the status of UPRH within the UPR system. There are other concrete examples of institutional change that will certainly persist beyond the funding period. The Human Resource Office was reluctant to release crucial necessary data for the ADVANCE program because of concern about confidentiality and lack of personnel to provide the data. Having been enlisted in different committees and activities, the office now realizes the significance of this data to the ADVANCE program as well as to the institution. Furthermore, the Statistics Division has agreed to incorporate in their work the data collection project that was put in place by the ADVANCE program. UPRH has a tradition of institutional planning and research and has made an effort to integrate planning and assessment processes. Therefore, analysis of institutional strengths and weaknesses in terms of recruitment, retention, and advancement of women faculty in science should now be incorporated in the institution's assessment and planning process. Well beyond an ADVANCE program and as part of their strategic planning, universities must examine the political, economic, social, and technological factors that create or remove barriers to the advancement of women faculty in science. In addition, institutions need an assessment-evaluation and action plan that monitors recruitment, retention, and advancement of women faculty in science within the institution's permanent planning office.

At the UPRH the Institutional Transformation award from the NSF ADVANCE program has had a positive impact on female faculty recruitment, retention, promotion, and self-image. As a result of this effort, five women science faculty were either hired in tenure-track or tenured positions in the following departments: Computational Mathematics (two), Social Sciences (one), Marine Biology (one), and Physics (one). The total faculty in UPRH science departments has remained constant at 104, as a system-wide faculty hiring freeze was in place during this period; however, women who were previously in temporary contracts or nontenured positions have had their appointments changed or they were promoted and tenured. Two of the five new faculty positions are replacements of recently retired women faculty within the Department of Computational Mathematics and represent no net change in the number of women science faculty members.

From Discretionary Hiring Process to Institutional Policies on Faculty Searches: Changing Policies amd Practices from the Department through the System Levels

Another key element for the advancement of women faculty is the recruitment process (Massachusetts Institute of Technology 1999). With the support of a legal advisor, processes and practices of faculty recruitment and promotion in each department have been identified. This strategy is based on a study to identify the needs and problems experienced in implementing measures that widen opportunities for women. Research findings indicate that although some departments conduct effective searches to recruit qualified and diverse faculty, the lack of formality and consistency makes them the exception rather than the rule. At UPRH, no consistency is found among departments about which person or committee is responsible for carrying out searches. The study also found no uniform strategy for creating an applicant pool or standard mechanisms to publicize open positions. Gender diversity has not been taken into account, and some departments have limited consideration of candidates to those working as non-tenure-track professors or to those referred by established graduate programs in Puerto Rico. Most recruitment processes are based on the belief that mastery of Spanish is an indispensable requisite. As a result, the application pool is significantly reduced or even limited to those known personally by current faculty members.

The project has also identified the need for an institutional policy on dual-career hires (Massachusetts Institute of Technology 1999). Hiring mathematics and science faculty with terminal degrees at the UPRH is difficult. Attracting faculty from the U.S. mainland is also a challenge. Providing opportunities for spousal hires would provide an incentive for faculty couples to relocate to the island. Despite the fact that the concept of dual-career hires was unknown, an ADVANCE institutional research project found that in 2004, 25% of the male and 35% of the female faculty in science were or had been part of a dual-career couple and that for years department chairs had made informal arrangements to hire couples. These practices have to be formalized and applied to all departments by transforming them into policies.

To develop the new procedures for faculty recruitment, ADVANCE UPRH has developed a faculty manual with detailed steps and best practices on how to recruit faculty. The manual focuses on gender diversity and guides UPRH to also focus on diversity in race, ethnicity, origin or social condition, national status, religion, and so forth. To develop these new procedures, ADVANCE UPRH facilitated dialogues with the department chairs and with the Coordinating Committee members. We decided to have as many discussions as necessary so that all key players would reach consensus and acknowledge the need for formal procedures. Discussions were coordinated by the legal advisor who previously researched all the university regulations and relevant literature. The legal advisor also conducted research based on individual interviews with each of the science department chairs, the human resources director, and the EEO (Equal Employment Opportunity) officer. She then developed a step-by-step guide for the Regulations Committee discussions.

The manual establishes a generic and uniform mechanism that clearly states each step of the recruitment process. The process begins with the department's planning committee identifying the needs and determining the target areas for the recruitment of faculty. When there is an opening for a faculty position, the department chair designates a search committee that includes at least one person committed to promoting women and other historically excluded segments of the population. Search committees have rarely been used at UPRH, but as a result of ADVANCE, they now have an important role. The search committee writes and publishes the job postings in collaboration with the EEO office. The manual establishes that announcing openings for faculty positions is mandatory and the administration will set an amount in the budget, in either the individual departments or at the level of the academic dean to cover the publication costs. All open positions are announced, including non-tenure-track full- and part-time positions. The search committee screens the candidate pool. Before starting the search, they define the process that will be used to evaluate the established requirements and criteria and ensure that evaluation of candidates is uniform regardless of gender, race, ethnicity, or other consideration. After each recruitment process, a report is submitted to the EEO, the Human Resources Office, and the chancellor, in which the recruiting process and each of its stages is detailed, as well as the efforts made to attract women candidates and other segments of the population identified in the UPRH Affirmative Action Plan. Materials are provided to assist and guide the process. Among other items included are examples of effective job postings for positions, a list of journals, professional associations and other sites consulted by women and other persons from the groups that the campus is interested in attracting, and letters to send the candidates and others that the chairs and search committees can use as templates.

At the same time that these policies were being developed, ADVANCE has worked closely with the Physics and Electronics and Computational Mathematics departments on their recruitment processes. Particular attention was given to the language used in the ads and to where they are published. As a result, 20% and 27%, respectively, of the applications in the two searches in these departments were from women, compared to 0% in the previous search. Highly qualified women candidates applied, and the Physics department hired its second woman faculty member, the first faculty member to be recruited with full professor rank. This successful example is included in an intensive educational campaign to promote the new regulations on campus.

Training Undergraduate Women Science Students for Academic Careers

A unique element of our ADVANCE program is the Faculty in Training effort. This is a program whereby faculty identify promising undergraduates and train them intensively in the research process and in the graduate admissions process, with the goal that the young women will earn Ph.D.s on the mainland and return to Humacao as faculty. Historically, UPRH has faced difficulties recruiting faculty in sciences and technology, particularly in the Applied Mathematics and Physics programs. In addition to the problems common to other higher-learning institutions in the United States, geographical isolation, lower salaries, and Spanish as the language of instruction make the recruitment more difficult. Our experience during the last decade is that after receiving appropriate mentoring, gifted students with clear academic inclinations have finished their Ph.D.s and returned as faculty (Negrón and Ramos 1999). With this experience in hand we set up a mentoring and instructional program for a selected group of women undergraduates: the Faculty in Training (FIT) program. The institutional research carried out as part of the ADVANCE program corroborated that this is a productive approach and that institutional and faculty support is an important factor for women who pursue graduate studies (Cruz Pizarro and Benítez 2003). Data were collected using group interviews and questionnaires administered to samples of students enrolled in science undergraduate programs and to alumni who had graduated with the highest qualifications and had potential to pursue graduate studies. The results of the study showed that female students who went on to graduate school based their decision on the following factors: confidence in having the necessary skills and ability, knowing their interests, knowing the admission process, and the support they received from their family. On the other hand, a reason not to pursue graduate is lack of financial resources. Our experience is that having adequate information about scholarships and learning how to apply to graduate schools is part of viewing the admission process as surmountable. All of these aspects are developed through the FIT program (Ramos, Vega, and Valdés 2003).

Interestingly, having to relocate outside of Puerto Rico for graduate school was not mentioned explicitly as a main obstacle, although we know it is one of the major concerns of our students. Students who were interviewed expressed this concern as a family issue when they said that they did not want to or could not move from Puerto Rico. They expressed feelings such as, "I have family responsibilities," or "I plan to marry." These statements are in accordance with the role families still assign to Hispanic women.

The FIT participants are recruited based on their academic record, recommendations from faculty, and area of interest for graduate studies. The students are advised to select an area that matches their interests and the needs of the institution and are encouraged to pursue studies in multidisciplinary programs to increase their possibilities of obtaining a faculty position after graduation. As part of the program, we require participation in undergraduate research, preferably with a faculty participant in our ADVANCE Institutional Transformation program. The institutional research confirmed our perception of undergraduate research as being a key element for the FIT component. Students who participate in undergraduate research not only have the opportunity to work in a lab for either a stipend or a grade, but they also acquire problem-solving skills, learn to make presentations, participate in conferences and seminars, and write scientific papers. They also learn one of the key elements of succeeding in academia: establishing a close relation with a mentor. In our experience there is no better way of building the confidence of a student in her capacity to pursue graduate studies than of having a paper accepted for publication. During the school year FIT participants attend leadership workshops, teaching practice sessions, and seminars offered by experienced male and female faculty. We bring female role models from academia, industry, and government to discuss leadership, education, and employment. A society (network) of participants, mentors, and role models has been established to encourage their continuous commitment to research and leadership as well as discuss pertinent societal issues (equality, violence against women, and others). Socialization within this network has proven to be vital for their continuous participation. Once the participants reach their senior year, we fully support their transition to graduate school by providing GRE training, language laboratories, and training on navigating graduate schools' admission protocols.

While specific workshops have been designed on how to work with family issues, we still have to manage the issue case by case. Most of our students live with their parents and have to negotiate with them to stay longer in the lab, to travel to a conference, or even to be allowed to do research. Very often the mentor has to talk directly to the parents. We have found ways to deal with these issues and therefore invite family members to the presentations, which help them feel proud of their daughter's accomplishments. Nevertheless, one conclusion is clear: if we want to have more Puerto Rican women with advanced degrees in science, we cannot ignore the role of the family.

After three years of implementation, five FIT participants are in graduate schools across the United States in the areas of Computational Mathematics, Environmental Engineering, Oceanography, and Materials Engineering. Three of them have passed their qualifying exams and are working on their dissertation research.

Although the benefit of FIT for the student participants and for the future of the science departments is clear, the benefit for the faculty who mentor and train these students may not initially have been obvious. In a nonstructured group interview with FIT mentors, we asked them to talk about the benefits of the program for women faculty. All the mentors agreed that although mentoring undergraduate research students is something they do as part of their regular duties, what makes this experience different is the special qualities of the students. Among faculty's remarks are, "These students are not only talented academically but extremely motivated, self-driven and are not afraid of working hard," "She is the best student I ever had," and "She helped me increase my research productivity." Having a FIT student has also meant additional support from ADVANCE in mentoring them through the admission process to graduate school. The FIT experience has also helped faculty mentors to develop more effective mechanisms for mentoring undergraduates and women in science. For the only male mentor, the experience led him to research and to seek more information on women in science. All the mentors said that they would feel extremely satisfied if their former students returned to UPR as faculty, and that they would make important contributions to the institution. However, there is a concern that in some departments faculty members may feel threatened by employing former FIT students. Some strategies to minimize this potential problem were assessed and are being implemented. For example, to facilitate their transition to faculty positions, former FIT students are being invited to visit the departments to give seminars and meet with faculty and students. We are also keeping them informed on departments' recruitment plans and advising them on career decisions (courses, thesis research, postdoctoral positions) that are compatible with these plans.

Conclusions

UPRH's ADVANCE can serve as a model to advance women faculty in other undergraduate, small, or minority institutions. The following specific strategies can be used to implement similar programs in other institutions.

1. Take advantage of the work that has already been accomplished. Identify people and organizations that have been working with gender issues and involved them through a coordinating committee or other organization.

2. Identify key players from all levels of the institution. Collaborate and take advantage of supportive top administrators but do not rely only on them. Organizing people from the base (at the department level) requires more effort but will outlast administrative changes.

3. Women faculty in small and underfunded institutions can benefit from collaborative research, making the best use of shared resources and working in a more supportive environment. Support for research must include release time if the teaching load is high.

4. Minority and undergraduate institutions have access to a tremendous pool of talented female students. With proper motivation and mentoring these students can be trained to pursue academic careers. Understanding and acknowledging their family-related issues is a key in their success.

5. In developing policies, support from a recognized independent legal advisor who understands the academic and intellectual dynamics of the university and is informed on problems that affect women in academia can increase understanding of the issues and administrative approval and support.

NOTES

The authors are grateful to all our collaborators in the UPRH ADVANCE program, including UPRH administrators, Coordinating Committee members, science department chairs, ADVANCE coordinators, program evaluators, legal advisor, Advisory Committee members, and especially to the following program participants: Milca Aponte, Isabel Cintrón, Denny Fernández, Luz Raquel Hernández, Mariam Hernández, María de L. Lara, Neliza León, Maite Mulero, Ileana Nieves, Margarita Ortíz, Deborah Parrilla, Ivelisse Rubio, Alexandra Valdés, Esther Vega, María Vega, Sylvia Vélez, and Natalya Zimbovskaya, all of whom embraced and supported our ADVANCE goals and vision.

This effort is possible because many persons and groups at UPRH have previously worked on women's issues and over the years have provided the university community with fundamental awareness of issues of gender: Women's Affairs Committee, Program to Prevent Violence against Women, Women in Physics Association (now Feminist Laboratory), Faculty Association, and the Non-Teaching Workers Union. From the beginning, Jorge J. Santiago-Avilés from the University of Pennsylvania believed the work we were doing was important and encouraged us to submit the proposal; he continues to collaborate with and support the program. The former UPRH chancellor, Roberto Marrero-Corletto, demonstrated his leadership and academic vision in his support of ADVANCE and in other efforts to promote women in our institution.

We also want to acknowledge the National Science Foundation for financial support and for its commitment to diversity in science.

1. See Sturm's discussion on the role of organizational catalysts in her chapter "Gender Equity as Institutional Transformation: The Pivotal Role of 'Organizational Catalysts,'" in this volume.

2. Daniel Altschuler, a senior research associate and former director of the Arecibo Radio telescope in Puerto Rico; Helen Davies, a microbiology professor from the University of Pennsylvania, former AWIS president, and lifelong advocate for women in science; María D. Fernós, Puerto Rico, women's advocate; Angela Ginorio, associate professor of women's studies, University of Washington; María Pennock-Román, MPR-Psychometric and Statistical Research Consulting; and Janice Petrovich, director of the Education, Knowledge, and Religion Education, Media, Arts, and Culture Program of the Ford Foundation.

REFERENCES

CAWMSET. 2000. *Land of plenty: diversity as America's competitive edge in science, engineering and technology.* Washington, DC: Congressional Commission on the Advancement of Women and Minorities in Science, Engineering and Technology Development.

Cruz Pizarro, L. S., and S. Benítez. 2003. La extraordinaria participación de las mujeres en los estudios subgraduados en las áreas de ciencias en la UPRH. *4to Coloquio Nacional sobre las Mujeres.* Paper presentation and published in proceedings of the conference (4th National Colloquium on Women). Held at UPR Humacao, April 10–13, 2003. Conference organized by "Congreso Universitario para Asuntos de las Mujeres y los Generos" (University Congress for Women & Gender Issues): 1:63–66.

Ginorio, A. 1995. *Warming the climate for women in academic science.* Washington, DC: Association of America Colleges and Universities.

Massachusetts Institute of Technology. 1999. A study on the status of women faculty in science at MIT. *MIT Faculty Newsletter* 9 (4). http://web/mit/edu/fn1/women//women.html.

National Science Foundation. 2000. *Women, minorities, and persons with disabilities: 2000.* NSF 00–327. Arlington, VA: National Science Foundation.

Negrón D., and I. Ramos. 1999. Increasing the quality of physics and technology programs at UPR-Humacao by promoting gender diversity. Proceedings IEEE FIE, 13b5–7, San Juan, PR.

University of Puerto Rico at Humacao Planning Office. Statistics on students and faculty, Humacao PR, 1997–2004.

Ramos I., B. Hernández, and E. Díaz. 2003. The ADVANCE Program at the UPR Humacao, Proceedings 2003, WEPAN, Chicago.

Ramos I., M. Vega, and A. Valdés. 2003. *FIT (Faculty in Training): Cultivating women undergraduates for professional careers.* 33rd ASEE/IEEE FIE F1D-1, Boulder, CO. Paper presented at 33rd Frontiers in Education Conference. Organized by American Association for Engineering Education and Institute of Electrical and Electronics Engineers.

Valian, V. 1999. *Why so slow? The advancement of women.* Cambridge: MIT Press.

SIXTEEN

Gender Equity as Institutional Transformation

THE PIVOTAL ROLE OF "ORGANIZATIONAL CATALYSTS"

Susan Sturm

RESEARCH SHOWS that the "glass ceiling" in academia is kept in place by everyday interactions occurring across the entire spectrum of faculty life. At each step of the continuum from graduate student to full professor, women face small differences in treatment, and these small disadvantages accumulate to produce large disparities in status and opportunity (Valian 1999; Cole and Singer 1991). These differences in treatment often occur without anyone noticing. They reflect unconscious biases reinforced by cultural patterns and shared by men and women alike. Within highly informal, unexamined, and poorly managed decision-making processes, these biases operate unchecked at many pivotal points of academic advancement (Bauer and Baltes 2002; Etzkowitz, Kemelgor, and Uzzi 2000). Women also face structural barriers to full participation, such as work-family policies and underinclusive indicators of academic promise. Organizational culture and processes preserve these exclusionary policies, without ever inviting scrutiny of their validity (Trower 2004).

Women's full participation in the academy cannot be achieved without examining these multilevel decisions, cultural norms, and underlying structures (Ely and Meyerson 2000). Change thus requires a process of institutional mindfulness. This means enabling careful attention to decisions that ultimately determine whether women and men of all races will have the opportunity to thrive, succeed, and advance. Research shows that self-consciousness about the processes, criteria, and justifications for hiring and promotion decision making minimizes the expression of cognitive bias (Bielby 2000; Fiske 2004). Institutional mindfulness also requires the capacity to institutionalize ongoing learn-ing—learning about problems revealed by examining patterns of deci-sion making over time, as well as about creative ways of addressing those problems, advancing participation and improving academic quality. Finally, it entails building incentives for improving inclusiveness and achieving excellence into ongoing governance systems (Sturm 2001).

Universities' decentralized governance systems complicate efforts to achieve institutional mindfulness. Power is highly distributed in acade-mia, and change is often difficult to achieve (Birnbaum 1988; Trower 2004). However, the building blocks of systemic change are present in many institutions. There are influential faculty concerned about gender and racial inequality who are in positions to press for change but lack adequate occasions or incentives to do so. There are research and advo-cacy institutions with accumulated knowledge and networking relation-ships but limited access to those in positions of power. There are con-cerned insiders in positions of administrative responsibility who see the relationship of gender to deeper institutional efficacy and legitimacy but feel powerless to change the status quo. There is a substantial and acces-sible knowledge base about the status of women, the causes of women's underparticipation, and the strategies enabling women to succeed, but that specialized knowledge does not reach everyday decision-makers. There are underutilized professional networks that could bring faculty from different fields and institutions together on a regular basis.

Systemic change occurs through connecting the knowledge and action of these engaged participants. Changing institutional climates requires everyday leadership in the various nodes of decision making that determine faculty advancement (Meyerson 2001). Interventions to increase gender equity can design roles and processes self-consciously intended to produce these transformative synergies. This institutional approach lies at the core of a recent National Science Foundation initia-tive called ADVANCE. ADVANCE provides institutions with "institu-tional transformation grants" designed to increase women's participation in academic science by engaging universities in a process of institutional self-analysis, cross-institutional learning, and change.

A key aspect of ADVANCE's strategy involves the development of a new role that has proven to be pivotal in enabling systemic change. NSF has parlayed the familiar position of principal investigators into a role that places individuals with knowledge, influence, and credibility in positions where they can mobilize institutional change. Because of their core function of mobilizing change at the intersection of different systems, we have called these individuals "organizational catalysts." Organizational catalysts are individuals who operate at the convergence of different domains and levels of activity. Their role involves connecting and leveraging knowledge, ongoing strategic relationships and collaborations, and forms of accountability across systems. The role is not unique to. ADVANCE; organizational catalysts can be found in many settings. ADVANCE, however, places them at the center of its implementation strategy and builds their knowledge and ability to mobilize others into the role.

This chapter uses a case study of a large research institution with an ADVANCE grant to analyze the role of organizational catalysts as change agents and to consider the role's applicability in other institutional settings. It first introduces NSF ADVANCE's institutional transformation approach and the "organizational catalyst" role as a crucial component of the dynamic enabling institutional change. It then examines how organizational catalysts are playing a central role within the University of Michigan ADVANCE initiative—what they do, who they are, how they do their work, and why they are so important to an effective institutional transformation strategy. Finally, the chapter considers the implications of the organizational catalyst role beyond ADVANCE.

NSF ADVANCE as an Institutional Intermediary

Prompted in part by the 1999 MIT report documenting women's marginalization and underparticipation (Massachusetts Institute of Technology 1999), NSF undertook an analysis of its gender programs and determined that its individual grant strategy was not making a dent in the problem. This analysis led NSF to adopt ADVANCE—a foundation-wide effort "to catalyze change that will transform academic environments in ways that enhance the participation and advancement of women in science" (National Science Foundation 2001).

NSF ADVANCE does not prescribe specific programs, strategies, or outcomes. It instead promotes a methodology for strategically connecting knowledge and action to address identified problems. NSF does this through funding proposals for institutional transformation that articulate a conceptual framework, build on existing knowledge, present a workable plan based on analysis of institutional data, and document an effective implementation strategy (National Science Foundation 2001, 2005).

It then insists on ongoing monitoring and assessment of these interventions, relying heavily on collaboration and information sharing among ADVANCE institutions, along with peer review. It extends its impact to the larger field by requiring grantees to make much of their learning publicly available, by infiltrating the knowledge networks that universities operate within, by connecting ADVANCE grantees with their counterparts in other institutions, and by encouraging participation of key institutions in their programs.

The University of Michigan is one of nineteen institutions receiving institutional transformation grants in the first two rounds of funding. Through ADVANCE, Michigan has developed an integrated series of individual, departmental, and campus-wide initiatives. Individual initiatives include faculty career advising, research funds providing support for key transitions, and networks supporting women scientists and engineers. Departmental initiatives support departments aiming to improve their climates through departmental transformation grants and self-studies. Campus-wide initiatives include interactive theater interventions and a program called Strategies and Tactics for Recruiting to Improve Diversity and Excellence (STRIDE).[1]

> This committee provides information and advice about practices that will maximize the likelihood that well-qualified female and minority candidates for faculty positions will be identified, and, if selected for offers, recruited, retained, and promoted at the University of Michigan. The committee works with departments by meeting with chairs, faculty search committees, and other departmental leaders involved with recruitment and retention.[2]

Less discussed but perhaps even more significant, Michigan's ADVANCE grant institutionalized a structure that, from the outset, harnesses the knowledge and social capital of individuals with a track record for effective problem solving. ADVANCE enables universities to locate individuals with legitimacy, influence, and commitment to promoting change, and to equip them with resources, visibility, access, and legitimacy.

The principal investigator (PI) role, which NSF builds into its award process, is the linchpin in the development of this institutional design. Like the conventional principal investigator, ADVANCE PIs collaborate with a research team to develop experiments in their institutions, analyze their effects, and report on them. They wield the responsibility, accountability, and legitimacy built into the PI status. But NSF recasts the PI role to take account of the systemic dimensions of the gender equity project. NSF ADVANCE reinvents the PI role as a research-based change agent within the institution. The next section examines this innovative and pivotal role as it has unfolded at the University of Michigan.

Introducing Organizational Catalysts: Connecting Domains, Discourses, and Knowledge

In 2002, Mel Hochster, a distinguished University of Michigan mathematician and member of the National Academy of Sciences, won the Margaret and Herman Sokol Faculty Award in the Sciences. One of the University's most prestigious honors, the award carried with it a widely attended public lecture—typically used as an opportunity to celebrate the recipient's eminence and to feature pathbreaking research. Hochster chose this occasion to speak to a room full of mostly male scientists and mathematicians about gender bias. Hochster's award lecture, entitled "Women in Mathematics: We've Come a Long Way—or Have We?" discussed the situation of women mathematicians and other women scientists, partly from a historical perspective and partly in terms of problems that exist today. He described "overwhelming evidence of gender bias in the evaluation of candidates and in many other contexts. Even when procedures seem to be objective and fair, studies have shown that gender bias is significant and pervasive." Hochster's speech was described by many as an important turning point in the institution. In the words of one high-level administrator involved in gender equity at Michigan:

> People walked out of that meeting like they'd been thunder-struck. "I had never thought about this gender thing before . . ." It was that he, who was a member of the National Academy of Science, gave this talk. . . . It was the drama of his gesture that really affected people. The information had been out, and he just had such a huge impact. Why? The National Academy of Science gets it. He gives over this important occasion for himself. Instead of talking about math, he talked about the problem of gender in science. It was hugely important—an amazing lesson in how this progresses.

How did this prominent mathematician become such an effective gender mobilizer? Hochster was energized by becoming part of STRIDE—a group of scientists who used the methodology of scientific research and data to educate themselves and then others about the dynamics, causes, and possible remedies for subtle gender bias. His speech dramatically illustrates the power of placing individuals with social and intellectual capital in positions to mobilize learning and change. But Hochster did not become an "organizational catalyst" alone or by accident. His role resulted from the efforts of others playing a similar role, only on a broader scale. More specifically, the ADVANCE PI and steering committee developed STRIDE as part of a broader strategy to leverage the pedagogical capacities of strategically located individuals throughout the institution.

This section is based on a case study of the ADVANCE program at the University of Michigan in order to develop a theory about how organization catalysts are created and how they are able to motivate organizational change. The observations about organizational catalysts at Michigan should hold true elsewhere as well, allowing for variation in the specific details of their roles within their own institutions. To conduct the study, a research team interviewed faculty, department chairs, deans, administrators, STRIDE members, and participants in ADVANCE at Michigan. They were asked to describe their experience with ADVANCE, including key turning points, and the programs and interventions that were most and least successful. There was a strong consensus that the role I have called "organizational catalysts," particularly the principal investigators, steering committee, and STRIDE, were the linchpin of ADVANCE. For many, these organizational catalysts were the most important factors in what was perceived as ADVANCE's initial success. Most of those interviewed did see an improvement in search and hiring patterns, the culture of the institution, the involvement of women in positions of influence, and the overall academic environment. They viewed these changes as fragile and incomplete, but as dramatic nonetheless, particular compared to the impact of previous gender equity initiatives involving science and engineering. Although it is both premature and beyond the scope of this study to draw any conclusions about ADVANCE's impact, independent interim assessments confirm that the program has increased the hiring of faculty women in the sciences, improved the institutional environment for women, and in the process, fostered improvements in overall policies and governance at the University.

NSF casts the PIs of the ADVANCE projects they fund as the con-
ceptualizers, planners, coordinators, conveners, and mobilizers of the
institutional transformation process. In the context of ADVANCE, the
planning stage is built into the grant application process. At Michigan,
a group of five individuals took responsibility for assembling individuals and institutions that had been involved in promoting gender and
racial equity, as well as leaders identified as crucial participants in an
effort to institutionalize change.[3] These five included the deans of the
three colleges employing the largest number of science and engineering
faculty and an associate provost (who serve as co-PIs), and the director
of the Institute for Research on Women and Gender, who serves as the
project's PI. All were already members of a university-wide Committee on Gender in Science and Engineering, and all held administrative
roles that would allow them to make an impact. These five became the
project's steering committee. They created working groups that
included advocacy group members, experts on gender and race, and
administrators willing to commit themselves to increasing women's
participation. The project's administrative staff also pulled together
studies documenting what was already known about the status of
women and people of color at the institution, and undertook additional
preliminary studies to provide information necessary to prepare a proposal. They canvassed the available research on the barriers to women's
advancement and effective strategies for addressing those barriers, as
well as the voluminous reports from other institutions that had conducted gender and racial analyses. They proposed an integrated strategy
aimed at transforming people's understanding of how gender operates,
and increasing departments' capacity to attract, retain and advance successful women in academic science (NSF at the University of Michigan
2002).

As organizational catalysts, the PI and other steering committee
members occupy a hybrid role, one that requires knowledge, legitimacy,
and social capital to get powerful people to the table, include the steering committee in decisions, and allow the steering committee to
influence their practices. Organizational catalysts must also be able to
instill hope and trust in groups that have become skeptical about the
possibility of change. The background and qualifications articulated by
ADVANCE and possessed by the role's occupants have a crucial part in
equipping them to undertake the steering committee's multiple duties.
They tend to be respected scholars with administrative experience
within the department or the university who are known for their com-

mitment to academic quality and equity. They often come into the position having played a significant role as a mentor to graduate students and junior faculty, and having worked with faculty and administrators at different levels within the University. They are highly respected faculty bringing considerable knowledge, administrative experience, working relationships, and professional legitimacy to their role as steering committee members.

The Tools and Strategies of Organizational Catalysts

I have analyzed the interviews and reports to identify the strategies accounting for the effectiveness of the PI, steering committee, and STRIDE members as catalysts of meaningful systemic change. This analysis reveals three such strategies: (1) mobilizing varied forms of knowledge to promote change, (2) developing collaborations in strategic locations, and (3) maintaining pressure and support for action.

Mobilizing Varied Forms of Knowledge to Enable Change

Organizational catalysts have access to many different methods and forms of information relevant to addressing gender issues. Social science research provides one key form of knowledge. As part of her researcher role, the PI conducts or oversees surveys and statistical studies documenting patterns in women's participation throughout academic life. Her long-standing institutional relationships and status as an NSF PI help the steering committee gain access to data that has previously been unavailable or difficult to obtain. Their knowledge and influence enables them to gather crucial information about the micro-level decisions that accumulate to shape access, such as data on offers, work assignments, research support, and the composition of the candidate pools actually considered in a search. They can then institutionalize this data-gathering so that reliable and relevant information is routinely produced. The PI and ADVANCE staff buttress their analysis of institutional data with climate and demographic studies from other institutions. They also collate and analyze the relevant scholarly literature on how gender bias operates in evaluations of men and women, and the types of interventions proven to reduce this bias. Equipped with this multifaceted knowledge, the steering committee then develops a conceptual framework to guide the institutional transformation project. According to one interviewee:

The strength of ADVANCE here is the bringing together of the social scientists and the scientists. Having someone with [the PI's] expertise as the leader of this and the scientists and engineers also deeply involved is important. We took the approach of study from a social science perspective. What Michigan is known for as an institution is social science research—with [Institute for Social Research] ISR, and what [the PI's] expertise is.

In addition to this empirical evidence, the steering committee's prior work within the institution, along with members' extensive interactions with different constituencies around issues of gender, provide them with cultural knowledge about the institution they seek to influence. The steering committee members often spoke of their familiarity with the history leading up to current conditions, coming out of their experience working on these issues over the years. They describe knowledge of where important decisions get made, who has influence within the department, and how people interact and advance. They have access to insider knowledge of what is valued and who has power, within particular departments. This informal knowledge equips them to work effectively within departments, to enlist allies, and to head off problems before they erupt into crises.

The steering committee members' work as troubleshooters and ombudsmen provides them with informal knowledge about the breakdowns or bottlenecks affecting women in particular departments. They learn about problems stemming from unsupportive managers, dysfunctional systems, or simple lack of awareness, and are in a position to intervene at the appropriate level within the university. Their work over time and across different departments also provides information about overarching problems that require coordinated or centralized interventions. For example, a committee focusing on recruitment, retention, leadership, and career development produced information about the impact of dual careers and work-family issues on recruitment and retention. The group identified the need for systemic change to address these problems that recur at the departmental level. The involvement of high-level administrators in the committee's ongoing work facilitated a successful process of policy change and implementation:

Some of the recommendations require university involvement, i.e., day care. Some of it is college level, some departmental. We have implemented a lot of these things. We will have a training manual about recruiting for search teams to talk about strategies and how to create a diverse pool and evaluating candidates. There are big issues with dual career that we can address because we're so big, but we needed formal mechanisms to make it easier to work across the college boundaries.

Steering committee members draw on their knowledge constellation to calibrate information's form and function to the context and problem at hand. They draw on empirical data to demonstrate the existence of the problem and examples of success to demonstrate the possibility of change. They analyze their informal interactions to determine the need for more systematic research. They also draw upon qualitative information gleaned from troubleshooting to help identify the source of gender disparities evident in the demographic data. Conversely, patterns revealed by the empirical research guide how and where to focus their problem solving interventions. The combination of methodologies permits strategic use of additional empirical research, based not only on whether the problem is well documented in the secondary literature but also on an assessment of what it will take to reach different constituencies.

The steering committee's combined responsibility for research and action may explain its extensive efforts to tailor the form of communication to particular contexts and disciplinary cultures. Members devote considerable attention to the question of how knowledge about the dynamics of gender bias can be effectively communicated to diverse (culturally, methodologically, and demographically) communities. They thus value social science research not only for what it teaches about the underlying problem, but also for its cultural authority. They proceed on the premise that data is only effective if it reaches the people who are in a position to act on that information. So the PI has observed that data must be communicated repeatedly and in many different forms.

The steering committee also uses knowledge to empower people to act, to legitimize the need for change, and to enlist the participation of key collaborators. Members enlist the help of the most effective communicators within particular settings. Social science data played a significant role in recruiting people to become active in ADVANCE. One STRIDE member described his reaction to the PI's presentation of social science evidence as a turning point in his decision to join STRIDE:

I said no initially . . . partly I was a little bit skeptical that a committee could do anything effective. . . .But after I heard her I
changed my mind and agreed to be on the committee. . . There
was a lot of information about climate at the U of M, and that
made me feel that the problem was larger than I had thought. I
think everyone on the STRIDE committee, as we studied the literature on gender bias, realized that the problems were larger than
people thought.

Every STRIDE member interviewed emphasized that their exposure
to the social science data also increased their capacity and willingness to
intervene about gender. Knowledge, in the currency of science with
data to support it, gave them tools, arguments, and confidence that they
otherwise didn't have. STRIDE members used the credibility of science
to identify gender bias as a serious problem justifying institutional
change:

They were data-driven, so it's incredibly convincing to skeptics.
In our department, people were open enough that they would
come out saying, "Wow, I didn't know that." We had them come
in again this fall, and required the search committees to be there.
A lot of what they do is provide data on evaluation bias. It
becomes a very scientific discussion about the evidence and the
nature of the evidence. People get engaged in the substance of it
as a scholarly issue. This was timed to take place directly before a
search. I had specifically talked to them about letters of recommendation, and the search committee read papers on this. . . .
People went back and started looking at their own letters.

The PI, along with STRIDE members, also learned through experience
that, for people to internalize the information, they required adequate
incentives to pay attention to it. One strategy the PI used to motivate
learning involved connecting the gender equity data to core concerns of
the department:

Another use for the data was to go into each department with a
picture of national and local data and have a one on one conversation with the chair. . . . To get the chair's attention, we would
figure out something that bothers them. Like graduate students
not going on to Ph.D.s or academic positions or attrition. It is

important to hook into something that is bothering them. . . . This provided a way to reach a department where not much or nothing is happening.

With experience, STRIDE shifted its focus to target the pivot points of decision and action and the individuals directly involved in those decisions, such as active searches or looming retention issues. This made STRIDE's information relevant, important, and immediately usable.

The steering committee members did not limit themselves to scientific modes of gathering and communicating knowledge. They developed other methods that could motivate interactions among faculty about issues that were never before recognized or discussed. One way they did this was through teaming up with a well-established teaching and research institute that used interactive theater to build knowledge:

> Using data from our interviews and from many studies nationally, they developed a sketch that presents a faculty meeting discussion of a recruitment. The sketch illustrates how a variety of non-conscious schemas and gender dynamics can lead a group to . . . less than optimal decision making about hiring and other matters.

The steering committee connected the CRLT Players (see chapter 13 in this volume) to deans and faculty, thus enabling a discussion of issues that must surface as part of a process of culture change. As one participant noted:

> Theater draws you in in a way that empirical data doesn't. There's an immediacy that you almost have to react to. It is when you get beyond resistance . . . and into the climate issues. People start talking about things in a way they haven't talked about it before.

The steering committee also participates in awarding funds designed to encourage departmental experimentation, and use the grant-making process to influence conduct and shape priorities within departments that choose to participate. These funds have supported departmental transformation efforts that operate like mini-NSFs located within their own institution, using funding to encourage experimentation and creativity. They provide support for innovative approaches to routine practices such as recruitment, selection processes, mentoring, and faculty support. The steering committee helps develop criteria for allocating these funds, offers technical assistance to applicants, and facilitates the process by which funding decisions are made.

Developing Collaborations in Strategic Locations

The steering committee and ADVANCE staff involved women faculty (along with their male colleagues) directly in the process of defining the focus of ADVANCE, in part as a way of mobilizing people to take action and creating a venue for women's voice. They have institutionalized this role of connecting women to each other as a way of mobilizing their knowledge and rekindling a sense of hope and possibility. This is one example of a second overarching function served by ADVANCE. It creates new "communities of practice" among individuals who share common interests, experiences, or concerns but otherwise lack opportunities to connect. The steering committee use its role to multiply occasions for women to meet, share their experiences, develop shared conceptual frameworks to inform their problem-solving strategies, and collaborate around issues of common concern. The steering committee's role in the formation of an informal network among the women science chairs offers one example:

> There are now five women chairs of science departments campus-wide. . . . So we decided, okay, five's a number. We could have a group. So we invited them to lunch. They all came. I said at the end of this, . . . you guys could meet on a regular basis and be a group. . . . We'll convene you, we'll schedule you, we'll make the reservation, we'll pay for lunch, but you don't need to have us there. . . . By the time they left, they wanted monthly meetings. . . . They were eager. They used the time, they came up with dilemmas they shared with each other and got advice from one another. It was great. So they're learning to do it. They are learning how to be a collective and how to define their own needs.

Other newly formed working relationships have put STRIDE committee members and others committed to women's advancement in regular contact with people in power around issues directly affecting women's advancement. One chair has worked very closely with a member of his department who is also on the executive committee and a member of

STRIDE. Over time, the chair describes how he has become more mindful as a result of those interactions.

> There are simple, commonsensical things that she keeps pointing out to me. We really need to make sure that we shouldn't have an admissions committee where there is not a woman on it. We shouldn't have a graduate committee which has advising responsibilities for students without women's participation. [The STRIDE member] is the one who is my conscience. Anything I start to do where I am not thinking, [she] points out and says, you ought to think about doing it differently. I say, whoops, you're right.

The ADVANCE steering committee members also meet regularly with chairs, deans, and other governance actors. These meetings provide regular occasions to connect gender issues to routine decisions. The steering committee creates new collaborations as well, bringing together groups that would otherwise never interact, to come up with solutions addressing common problems. They have developed task forces and committees to integrate new understandings about gender equity and organizational improvement into policy and administrative governance. They also identify faculty in a position to exercise moral leadership, and then equip them with the tools and support to speak up when they see a problem involving gender in the course of their daily routines. They thus bolster decisions to exercise everyday leadership at key pivot points defining access and participation. The architecture of the ADVANCE initiative increases the number of these pivot points and decreases the risk of taking action. These structural innovations sustain the conditions permitting activism to flourish and leadership to emerge (Meyerson 2001; Katzenstein 1990).

The steering committee members have also become national intermediaries of institutional change. They collaborate with their counterparts at other institutions, developing best practices, metrics of effectiveness, and toolkits for intervention that can be adapted to different institutions. They evaluate each other's programs, both informally and as site visitors and external evaluators. They are invited into institutions that are beginning the process of institutional change, where they speak publicly, share their knowledge with local leaders, and give feedback on proposed plans. They are also contributing to the field's development by writing in peer-reviewed journals and editing books, including the volume in which this chapter appears.

Creating Pressure and Support for Change

A third crucial role performed by organizational catalysts involves keeping the pressure on. The steering committee members have referred to themselves as burrs, nudges, "articulate pains in the ass," monitors, and prodders of change. They create occasions and incentives for people in positions of responsibility to act, and for people who care about gender to press for change. They maintain the institution's focus on gender as part of its core mission. They keep problems on the front burner and help put together workable solutions, making it harder not to take action. They see their role to require them to "hold the institution's feet to the fire and make sure that it gets institutionalized."

How do organizational catalysts do this? They spot gender issues when they come up and make sure they are the subject of explicit discussion. They put issues affecting women's participation on the agenda. They help create multiple constituencies for change—constituencies who otherwise wouldn't see their interests as overlapping. They frame issues so that faculty concerned about the quality of the graduate student experience and about faculty retention join with those concerned about the climate for women and people of color to push for change. They arrange meetings with high-level administrators so that they can hear the arguments from influential faculty together with advocates for improving the institution's involvement of women and people of color. They use the evidence from the data to demonstrate the existence of the problem and construct a case for action. They use their social capital and that of others whom they have brought into the process to make it more costly to do nothing. Perhaps most importantly, the organizational catalysts help figure out what to do, and then they do the legwork to maintain the momentum so that these proposed changes actually occur. Their sustained attention to the issue and their follow-through with concrete action plans makes it much easier for high level administrators to take action.

Moving Beyond ADVANCE

The organizational catalyst role has implications well beyond ADVANCE. If the role can be institutionalized and adapted to other settings, it could become a significant element of any change initiative requiring cultural or institutional transformation. Organizational catalysts could be the cornerstones for sustaining the change process begun by ADVANCE over the long run. Like other ADVANCE institutions, Michigan is exploring whether to create a permanent position within the university administration to sustain ongoing institutional transformation. Michigan has extended STRIDE to departments beyond the scope of ADVANCE, and has committed to continuing its operation. The organizational catalyst role has also surfaced in race and gender equity initiatives undertaken by institutions acting without NSF support. Some universities have created new administrative positions with responsibilities similar to ADVANCE PIs, such as Vice Provost for Diversity Initiatives at Columbia or the Senior Vice President for Diversity and Faculty Development at Harvard (Harvard University 2005).

Institutionalizing the organizational catalyst role holds considerable promise as a means of building in ongoing institutional mindfulness and accountability. There are, however, risks attached to relying upon a permanent organizational position as a change strategy. First, there is the risk of role substitution: reliance on an institutional position in lieu of a well-researched concept and action plan. Some non-ADVANCE institutions appear to have created a high-level position to spearhead a change process without supporting the institutional self-study and strategic planning so crucial to the role's effectiveness. These initiatives may also fail to incorporate monitoring and external accountability into the role's operation. Some internally generated proxy for NSF's grant application, monitoring, and renewal process might help to assure that the organizational catalyst role remains tethered to evidence-based planning and action.

Second, there is the risk of overcentralization. The position could foster the expectation that the responsibility for change lies primarily with this administrative official. The role-occupant might also be tempted to use a top-down strategy relying on formal administrative authority and access to push through policy changes. This approach would undercut the development of shared responsibility for change, and induce passivity by faculty and administrators whose active participation is necessary for cultural and systemic change. Overcentralization also encourages deference to administrative decisions, and limits the capacity of faculty to hold the organizational catalyst accountable for her actions. Centralization of responsibility in a single individual also renders the change initiative vulnerable if the occupant of the position were to leave. The organizational catalyst role could be structured to minimize these risks by allocating responsibilities among different people, creating participatory oversight by groups in a position to evaluate the work of the office, and requiring ongoing public reporting on the office's activities and impact.

Finally, there is the risk of bureaucratization. Part of what makes the organizational catalyst role work is its fluidity and experimental character. PIs and STRIDE are constantly reinventing themselves to respond to changes in the environment. If the position becomes too directly intertwined with and accountable to the central administration, it risks losing its independence, its openness to adaptation, and ultimately its legitimacy. If the position's occupants become full-time administrators for too long, they might lose scholarly credibility and access to local knowledge and thus also lose the social capital so crucial to the role's effectiveness. Over time, the role could become routinized and divorced from a change process with adequate resources and connections to constituencies for change, and, at worse, devolve into a symbolic or toothless position. An unlimited term in an administrative position may also blunt the sense of urgency and drive that the PIs now bring to their role. The relentless questioning of the status quo, which seem so crucial to the position's impact, may be difficult for one person to sustain over the long run, especially without a break.

The challenge is to define a long-term role that institutionalizes the experimental qualities of the organizational catalyst. The role's effectiveness depends upon cultivating the qualities that make NSF PIs and STRIDE members so effective: professional legitimacy, insider/outsider status, operation at the intersection of multiple systems, evidence-based decision making, deep knowledge of relevant contexts, and external accountability. This poses essentially an institutional design problem. The position could be structured to build in collaboration with diverse constituencies. Checks against co-optation and bureaucratization could be achieved by establishing rotating and shared positions, which might also make it easier to recruit high-status faculty for these roles. It is also important that these roles maintain independence from the central administration as well as accountability to constituencies committed to gender and racial equity, including peer institutions involved in similar work. Organizational catalysts could themselves be crucial participants in designing the expansion and institutionalization of the role, with their successors in mind.

Conclusion

This study of organizational catalysts, as conceived by NSF ADVANCE and implemented at the University of Michigan, shows the promise of this role innovation as a way to achieve the institutional mindfulness so crucial to full participation by women and people of color in the academy.

NOTES

The author is grateful to ADVANCE participants at the National Science Foundation and at the University of Michigan, who were so generous with their time and insight. This study would not have been possible without the outstanding research assistance of Kati Daffan.

1. http://www.umich.edu/~advproj/about.html. For an in-depth analysis of STRIDE, see Abigail J. Stewart, Janet E. Malley, and Danielle La Vaque-Manty, "Faculty Recruitment: Mobilizing Science and Engineering Faculty," in this volume.

2. http://sitemaker.umich.edu/advance/STRIDE.

3. Case studies and reports of other institutions suggest that organizational catalysts are playing an important role in other ADVANCE programs as well (Idalia Ramos and Sara Benítez, "Advancing Women Science Faculty in a Small Hispanic Undergraduate Institution," in this volume).

REFERENCES

Bauer, C. C., and B. B. Baltes. 2002. Reducing the effects of gender stereotypes on performance evaluations. *Sex Roles* 47 (9/10): 456–76.

Bielby, William T. 2000. Minimizing workplace gender and racial bias. *Contemporary Sociology* 29 (2): 120–29.

Birnbaum, Robert. 1988. *How colleges work: The cybernetics of academic organization and leadership.* San Francisco: Jossey-Bass.

Cole, J., and B. Singer. 1991. A theory of limited differences: Explaining the productivity puzzle in science. In *The outer circle: Women in the scientific community,* ed. H. Zukerman, J. R. Cole, & J. T. Bruer, 277–310. New Haven: Yale University Press.

Ely, Robin, and Debra Meyerson. 2000. Theories of gender in organizations: A new approach to organizational analysis and change. In *Research in Organizational Behavior,* vol. 22, ed. B. Staw and R. Sutton, 105–53. Amsterdam: Elsevier Science and Technology Books

Etzkowitz, H., C. Kemelgor, and B. Uzzi. 2000. *Athena unbound: The advancement of women in science and technology.* New York: Cambridge University Press.

Fiske, S. T. 2004. Intent and ordinary bias: Unintended thought and social motivation create casual prejudice. *Social Justice Research* 17 (2): 117–27.

Harvard University. 2005. Report of the Task Force on Women Faculty. http://www.news.harvard.edu/gazette/daily/2005/05/women-faculty.pdf

Katzenstein, Mary Fainsod. 1990. Feminism within American institutions: Unobtrusive mobilization in the 1980's. *Signs* 16:27–54.

Massachusetts Institute of Technology. 1999. A study on the status of women faculty in science at MIT. *MIT Faculty Newsletter* 9 (4). http://web/mit/edu/fn1/women//women.html.

Meyerson, Debra E. 2001. *Tempered radicals: How people use difference to inspire change at work.* Cambridge: Harvard Business School Press.

NSF ADVANCE at the University of Michigan. 2002. http://www.umich.edu/~advproj/proposal.pdf.

National Science Foundation. 2001. ADVANCE: Increasing the participation and advancement of women in academic science and engineering careers. Program announcement, nsf0169. http://www.nsf.gov/pubs/2001/nsf0169/nsf0169.htm.

———. 2005. ADVANCE: Increasing the participation and advancement of women in academic science and engineering careers. Program announcement) nsf05584. http://www.nsf.gov/pubs/2005/nsf05584/nsf05584.pdf:5.

Sturm, Susan. 2001. Second generation employment discrimination: A structural approach. *Columbia Law Review* 101:458–567.

Trower, Cathy. 2004. Advancing and evaluating impact. NSF ADVANCE National Conference, April 20. http://www.advance.gatech.edu/2004conf/3a_trower.ppt.

Valian, V. 1999. *Why so slow? The advancement of women.* Cambridge: MIT Press.

SEVENTEEN

Institutionalization, Sustainability, and Repeatability of ADVANCE for Institutional Transformation

Sue V. Rosser & Jean-Lou A. Chameau

EXPERIENCED INDIVIDUALS contemplating a project with goals of transforming the department, college, or university build sustainability and institutionalization into the initial planning of the project. For example, principal investigators sow the seeds for successful institutionalization and sustainability of their ADVANCE grants when they make the decision to submit the grant and plan the goals, objectives, and activities underpinning the particular aspects of institutional transformation that their university will pursue within a general framework to advance faculty women to senior and leadership positions. Receiving the National Science Foundation (NSF) funding in a very competitive, peer-reviewed program with relatively large grants carries considerable prestige. The institutional investment of both human and capital resources and commitment to establish, change, and implement policies and practices to support ADVANCE leverage the NSF support and assure long-term impact of the initiative.

Although institutionalization, sustainability, and repeatability sound like issues addressed during the final phases of a multiyear, several-million-dollar project, in fact they stand as some of the factors that need to be considered first, even prior to the initial, planning phases of any project. If the institution, and particularly the upper levels of institutional leadership, have not been informed, do not understand, or have not committed to sustaining the trajectory and impacts of a complex project such as advancing women to senior faculty and leadership positions after the grant funding ends or the individuals leading the project leave the institution, then the project will not succeed in sustaining this institutional transformation in the long run. Although the NSF grants carry considerable prestige, many institutions such as Princeton (Bartlett 2005) and Harvard (Pope 2005) undertake similar initiatives using institutional, rather than grant, funds. The institution's investment of resources and its commitment are the real keys to success. Even those institutions with NSF or other external support find that the institutional investment is as critical as the NSF support, even during the short term of foundation funding.

Summary of Georgia Tech's ADVANCE Initiative

The NSF ADVANCE grant serves as a case study from which we can outline our approach to institutionalization and sustainability of projects focused on institutional transformation. In order to understand the perspective from which this chapter comes, we provide a brief synopsis of the five major threads of Georgia Tech's ADVANCE project so that readers may evaluate how the particularities of this project might be generalized:

Thread 1. *Termed professorships to form a mentoring network:* One tenured woman full professor in each of four colleges with disciplines funded by NSF became the designated ADVANCE professor. The title and the funds of $60,000 per year for five years associated with the ADVANCE Professorship conferred the prestige and funds equivalent to those accrued by other endowed chairs at the institution. This sum also meant that $1.2 million of the $3.7 million grant went directly to support the ADVANCE Professors, consonant with the NSF notion that the ADVANCE grants should be substantial to recognize the importance of activities to build workforce infrastructure. Because Georgia Tech is a research university, the principal investigators (PIs) of the grant particularly recognized the necessity for ADVANCE Professors to sustain their research productivity while undertaking this mentoring role. ADVANCE Professors often used funds to pay for graduate students or postdocs to support their research.

Each ADVANCE Professor developed and nurtured mentoring networks for the women faculty in her college. The focus of the mentoring activities varied among the colleges, depending upon the numbers, ranks, and needs of the women. In the College of Engineering, a large college with about forty-two women out of four hundred tenure-track faculty, isolation constituted a primary issue in many units. The lunches arranged by the ADVANCE Professor with women faculty from the College provided an opportunity for them to meet women in other departments and develop social and professional networks. A popular professional networking opportunity included evaluation of the curriculum vitae of junior faculty by senior colleagues to assess their readiness for promotion and tenure or gaps that had to be addressed for successful promotion to the higher rank.

The ADVANCE Professor often helps to explain and mediate problematic issues with the chair and dean. In the smaller College of Computing, with eight of sixty women as tenure-track faculty, many of the women had young children, so many of the lunches and activities focused on explication of family friendly policies and strategies to balance career and family. In the College of Science, lunches and activities centered on grant-writing workshops and other means to establish successful laboratory research. In Ivan Allen College, where 40% of the tenure-track faculty are women, the ADVANCE Professor chose luncheon themes on publication and scholarly productivity. Although all four ADVANCE professors held luncheons and mentored individual women faculty, each focused the initial activities upon those issues she perceived as most problematic and critical for achieving tenure, promotion, and advancement to career success for the women in her particular college. By the fourth year of the grant, the professors evolved more cross-college activities, expanding programs and initiatives particularly successful in one college to women from all colleges on campus.

Thread 2. *Collection of data indicators like those in the MIT report:* To assess whether advancement of women really occurs during and after the institutional transformation undertaken through ADVANCE, data must be collected on indicators for comparison with baseline data. Georgia Tech proposed in its grant to collect data on eleven of the following twelve indicators that NSF eventually required all ADVANCE institutions to collect by gender: faculty appointment type, rank, tenure, promotion, years in rank, time at institution, administrative positions, professorships and chairs, membership in tenure and promotion committees, salaries, space, and start-up packages.

Thread 3. *Family-friendly policies and practices:* Recent studies document that balancing career and family constitutes the major difficulty for tenure-track women in general (Mason and Goulden 2004) and women science and engineering faculty in particular (Rosser 2004; Xie and Shaumann 2003). Competition between the biological clock and the tenure clock becomes a significant obstacle for women faculty who have delayed childbearing until they receive a tenure-track position. For women faculty in science and engineering, significant time away from their research makes it less likely they can successfully achieve tenure in a research institution. The dual-career situation becomes an additional complicating factor for women scientists and engineers, 62% of whom are married to men scientists and engineers (Sonnert and Holton 1995). (Given the dearth of women scientists and engineers, the reverse does not hold: relatively few men scientists and engineers are married to women scientists and engineers.) To facilitate the balancing of career and family, perceived overwhelmingly by women scientists and engineers, particularly those of younger ages, as the major issue (Rosser 2004), Georgia Tech instituted the following family-friendly policies and practices: stoppage of the tenure clock, active service–modified duties, lactation stations, and day care. The specific details of these policies can be accessed under Family and Work Policies at http://www.advance.gatech.edu.

Thread 4. *Miniretreats to facilitate access to decision makers and provide informal conversations and discussion on topics important to women faculty:* Research has demonstrated that women faculty tend to have less access and fewer opportunities than their male colleagues to speak with the decision makers and institutional leaders (Rosser 2004). Often this unintended discrimination and lack of access result from women's absence from informal and social gatherings. To insure access of tenure-track women faculty to the senior leadership of chairs, deans, provost, vice presidents, and president, the Georgia Tech ADVANCE grant organized two-day miniretreats during each year of the grant. Focused on topics of interest and concern to all faculty, such as case studies of promotion and tenure, training to remove subtle gender and racial bias in promotion and tenure decisions, and effective strategies in hiring dual-career couples, these retreats have provided opportunities for the tenure-track women faculty to interact with the institutional leadership and express their views on matters of mutual interest.

Thread 5. *Removal of subtle gender, racial, and other biases in promotion and tenure:* Close involvement with the promotion and tenure process provides insight into subtle ways in which unintended biases might influence decisions on promotion and tenure. For example, we have observed that in some cases when the tenure clock has stopped for a year for a valid reason such as childbirth, the clock appears not to stop in the heads of colleagues as they consider the individual for promotion and tenure. Colleagues seem simply to expect an additional year's worth of papers, talks, and productivity to be added.

To address this issue, the principal investigator, who was also the provost, appointed a Promotion and Tenure ADVANCE Committee (PTAC) to assess existing promotion and tenure processes; explore potential forms of bias, providing recommendations to mitigate them; and elevate awareness of both candidates and committees on expectations and best practices in tenure and promotion. After one year of studying the research documenting possible biases due to gender, race or ethnicity, ability status, as well as interdisciplinarity, the committee developed nine case studies with accompanying sample curriculum vitae. Each illustrated one or more issues or areas where possible bias might impact the promotion and tenure decision. After discussion of these case studies at a miniretreat, the refined versions served as the basis for an interactive web-based instrument, Awareness of Decisions in Evaluating Promotion and Tenure (ADEPT), designed by colleagues in the College of Computing. Individuals can use ADEPT to participate in a virtual promotion and tenure meeting, where depending upon their response, the meeting takes different directions and generates different outcomes in promotion and tenure. The web-based instrument, along with best practices from PTAC and resources on bias, is discussed in Chapter Eleven and can be accessed at www.adept.gatech.edu.

Questions to Consider for Institutionalization

After working with several institutions that were deciding whether pursuing an ADVANCE grant would be appropriate, we evolved several questions and suggestions to facilitate their decisions. Honest, serious answers to these questions will help institutional leaders and potential principal investigators determine their readiness to develop a project that, in terms of scope, timeliness, and changes in financial practices, will be sustained and institutionalized.

TABLE 1.   Questions to Consider

1. Do your goals and objectives for the project fit with your institutional strategic plan and other goals and priorities?
2. Do you have project leadership appropriate to the level of transformation sought within the institution?
3. Do your activities build on extant programs and include faculty leaders in the program leadership?
4 Is your institution willing to take an open and serious look at its policies and practices at a variety of levels with the understanding that it must correct and change those that do not support the project goals?
5. Is the institution willing and able to contribute significant resources to this project?
6. If the project received external funds, where will the home for the project be when funding ends and how will the project continue?
7. Will the institution undertake the project, or at least a scaled-down version of it, even if external funding is not obtained?

1. Do your goals and objectives for the project fit with your institutional strategic plan and other goals and priorities?

Although most institutions include a commitment to increasing diversity as part of their institutional plans and priorities, the interpretation of diversity and commitment to addressing it for senior and tenure-track women faculty vary considerably among institutions. Many institutions focus on student diversity, but have different goals for faculty. For example, at women's colleges for many years the ideal has been to have approximately 50% of the faculty be men. Although some institutions have a goal of "working toward a faculty that reflects the diversity of the student body," the current reality is that women constitute 35% of faculty overall, including lecturers and instructors (Commission on Professionals in Science and Technology 2000), with the vast majority holding positions at the lowest ranks in the less prestigious institutions and in disciplines outside of science, technology, engineering, and mathematics (STEM). At many institutions, diversity equates with racial or ethnic diversity and fails specifically to include women. Although women now constitute 56% of undergraduate students in the United States (U.S. Department of Education 2004), no one suggests that women should constitute the majority of faculty. In contrast, some institutions have begun to worry about having too many women students (Fasbach 2005; Hong 2004). Some systems and institutions have instigated programs to attract male students, especially African American and Latino men (Pendred 2004).

What are the implications of these institutional diversity programs and priorities for success with an ADVANCE project? If increasing and advancing women is not an institutional priority, then the message of ADVANCE does not mesh easily. Most projects focus on advancing tenure-track women in science, technology, engineering, and mathematics to senior faculty, chair, dean, or other leadership positions (thread 2). In most STEM disciplines women remain an underrepresented group at all levels, from undergraduate students, through graduate students and postdoctorates to faculty. This dearth at all levels makes it relatively easy to convey the ADVANCE message of increasing the number and percentage of women tenure-track faculty in senior and leadership positions in a technologically focused institution such as Georgia Tech, because the ADVANCE message falls within the broader institutional message generally conveyed, that is, the desire to increase the number of women.

In institutions that are not technologically focused, the ADVANCE message is likely to fall into a different, mixed set of institutional goals and messages about women. For example, many comprehensive universities seek to increase the number of women students and faculty in STEM fields, while they want to increase the number of men students, and possibly faculty, in other fields such as education, library science, and nursing. Hence, the ADVANCE message falls into a mixed set of messages coming from the institution about women, which may vary by the particular discipline and differ for faculty and students.

In institutions with explicit goals and programs to attract more male students or with worries that their student body "is trending towards too many women or becoming feminized," the ADVANCE project falls into an even more complex institutional priority environment. At the least, someone may ask how this program for women fits with the goal of attracting more men students. To prevent backlash, the case for advancing tenure-track women into senior STEM positions must be articulated clearly and repeatedly, without appearing to contradict the particular institutional goal of attracting more male students. In sum, depending upon the institutional type and gender composition, insuring that the goals and objectives of the ADVANCE grant fit the institutional goals and priorities requires more than simply checking that the institution's strategic plan includes increasing diversity as a goal. A high degree of correspondence between project and university goals increases the probability of ultimate institutionalization of the project.

2. Do you have project leadership appropriate to the level of transformation sought within the institution?

An early decision that dictates much of the later planning for the grant centers on the level of transformation sought. Does the project seek transformation at the level of department, college, or the entire institution? The level determines the rank of administrative leadership that must be involved in the project.

For example, if one or two very large departments constitute the primary target for transformation, then the department chairs should probably serve as PI or co-PIs on the grant. Although the dean and provost might be involved in another capacity, such as serving on an internal or oversight advisory board, they need not have a key role in project administration. If the college serves as the major site for transformation, then the dean must have a key position in the grant itself. If the goal is transformation of a key component, such as tenure and promotion or hiring, applicable to more than one college or even the entire university, then the provost or chief academic officer needs to serve in the capacity of PI. Because a major thread (thread 5) of the Georgia Tech ADVANCE grant centered on removing subtle bias in promotion and tenure processes at all levels in all colleges, the provost needed to serve as PI of the grant to insure this level of institutional academic transformation.

Even though they have the respect of their colleagues and goodwill and approval of their department chairs, deans, and provost, faculty members serving as co-PIs on an ADVANCE project rarely hold positions of leadership in the formal, institutional hierarchy that would enable them to implement policies, procedures, and practices and to hire, promote, and fire. They are unlikely to be successful in bringing about transformation at the level of the department, and certainly not at the level of the college or entire university. In contrast, having the provost serve as PI on a grant where the department is the level of transformation sought may not be the best use of his or her time and is likely to result in the perception that the upper administration is driving the project from the top down. Some institutional cultures resist top-down implementation as contrary to faculty governance. Although the PI will need to communicate with levels both above and below those targeted for transformation, leadership that holds the position appropriate to implement transformation at the level sought increases the likelihood of success in an ADVANCE project.

Furthermore, the institution should critically assess and identify the appropriate target(s) for transformation and associated leadership. As the substantial literature on institutionalization of reforms in higher education documents, effective institutionalization must include the top leadership (Eckel and Kezar 2003; Heifetz and Laurie 1997), middle administrators (Meyerson 2003), and the faculty (Merton et al. 2004; Woodbury and Gess-Newsome 2002). Changes diffuse throughout the organization (Rogers 2003; Strang and Soule 1998) at different rates but must ultimately penetrate the structure, procedures, and cultural levels (Braxton, Luckey, and Helland 2002) of the university for genuine institutionalization. To succeed, the NSF ADVANCE program for *institutional* transformation requires a deliberate and clear match between the goals and values of an ADVANCE project and the actual needs of the institution for transformation. Although it may still result in a prestigious grant, gaining project funding in the wrong departmental or organizational structure will not lead to transformation of the institution as a whole.

3. Do your activities build on extant programs and include faculty leaders in the program leadership?

Having top administrative leaders also holding leadership positions complemented by significant involvement of women faculty leaders helps insure success of a project that intervenes on behalf of women. For example, on most campuses, a group of women faculty, often including research faculty and staff, have worked for years on issues closely related to ADVANCE. They have collected data on the status of women, built programs to advance women's careers, and worked to establish women's centers and studies programs. Typically they have built an infrastructure and network with limited help from foundations or the institution.

Most successful projects build on an existing infrastructure and leverage the expertise of well-established, respected campus leaders historically involved with the project's topics. Although the idea and drive behind the proposal and the PI for the project may not come from this group, an entirely new infrastructure and leadership for the project, separate from extant faculty initiatives, creates a formula for failure. For example, not involving respected faculty leaders on women's issues and STEM in the ADVANCE project signals to both faculty and administrators a lack of grassroots support and a gap between the project's leaders and appropriate communities on campus. It also indicates the project leaders will fail to benefit from years of experience in what works and does not work in transforming the institution. For example, the Georgia Tech project leaders include both respected senior women from STEM as ADVANCE Professors (thread 1) and the leaders of Women, Science, and Technology (WST), Georgia Tech's version of women's studies.

Often the leadership for women's issues on campus has not come from faculty in the STEM disciplines. Directors of women's studies and much of the scholarship on women emerged initially from the humanities (Boxer 1998; Garrad 2002), later from the social sciences, and only more recently from the sciences (Fausto-Sterling 1992; Rosser 1988, 2000). This dearth may result from the precise problem that ADVANCE seeks to solve—small numbers of tenure-track faculty women in senior and leadership positions in STEM. Women faculty in STEM often have concentrated their efforts in women in science and engineering (WISE) programs, which sometimes remain somewhat separated from other campus efforts, particularly if attracting and retaining women students are foci of the WISE program.

On a campus without a WISE program or women STEM faculty who are involved in women's issues, the engagement of key women faculty such as the director of a women's studies program, the head of the committee on the status of women faculty and staff, or president of the association for women faculty in leadership positions in the ADVANCE project will serve as a crucial link to both the administration and the faculty. Although some issues for women faculty in STEM are unique because of small numbers, long hours required for laboratory and field work, larger start-up packages, and necessity of grant funding to undertake research, other issues are shared by many women faculty. Faculty from the humanities and social sciences who have worked on policies and practices surrounding pay equity, balancing work and family, and elimination of bias in faculty searches, promotion and tenure can provide important insights about institutional transformation to the ADVANCE project team. Involvement of these key non-STEM leaders not only will help ADVANCE, but will also position the project as one that builds on existing efforts for women faculty and attempts a further level of institutional transformation.

4. Is your institution willing to take an open and serious look at its policies and practices at a variety of levels with the understanding that it must correct and change those that do not support the project's goals?

The explanation of questions 2 and 3 places heavy emphasis on appropriate leadership in the project from both the senior administration and faculty. Having both upper administration and grassroots support becomes critical, in different ways, for identifying and changing policies and practices to support the advancement of tenure-track women faculty for an ADVANCE project.

Women faculty have the experience, and in some cases the professional expertise, to identify the policies and practices that impede their career advancement. Family-friendly policies (see thread 3) such as stopping the tenure clock and active service-modified duties (Cook 2001; Mason and Goulden 2004) that help to alleviate competition between the biological clock and the tenure clock stand as examples of changes that lower some barriers to career advancement for women. Anecdotal information suggests that some women fear stigmatization for taking advantage of such policies. To overcome this concern, Princeton University is giving new faculty parents of both sexes automatic one-year extensions on the tenure track, rather than placing the onus on the faculty member to ask for the extension (Bartlett 2005). On-campus child care facilities and lactation stations also help in balancing career and family. The role of faculty in identifying such problems as eligibility (can both men and women stop the tenure clock? adoptive parents? non-married partners?), providing data to support the need for a change, and suggesting possible solutions, and the role of the administration in obtaining the necessary resources, space, and changes in institutional or system policies to implement solutions seem relatively obvious.

The areas of data collection and access to information complicate the roles, problems, and solutions. Collection and open access to data disaggregated by gender (see thread 2), rank, and discipline, such as information on space, start-up packages, salary supplements from grants and medical practice plans, graduate student and research support, rates and times to promotion, not typically accessible even in public institutions, but requested by the NSF ADVANCE program, may be problematic. Assessing and revealing such data may provide evidence that could be read as discrepancies or discrimination on the basis of gender. For that reason the upper administration of some institutions may be reluctant to collect such data, fearing that availability of the data might lead to conflicts on campus and even be used in litigation against the institution.

The individuals planning on proposing institutional transformation projects need to make the highest levels of the administration (deans, provost, president) aware that such data collection will be conducted as part of the research. Most upper-level administrators will recognize immediately that committing to the collection of such data is the right thing to do, but also implies that they will need to correct and remediate any substantiated patterns that affect the advancement of women (or other faculty groups) negatively. Deans, provosts, and presidents who agree to collect such data have demonstrated their good intentions and commitment to change policies and procedures that discriminate on the basis of gender.

Both the administrative leaders and the faculty leaders of the projects like ADVANCE must play their appropriate roles in communicating the reasons for the data collection and the commitment to correction, if necessary. For example, the provost may need to explain to the deans the reason for collecting data on hiring, promotion, tenure, retention, space, start-up packages, and salary supplements on individual faculty in the college. The administration will probably need to be prepared to develop a plan, including budgetary resources, for how the institution will rectify documented patterns of discrimination. The provost might reassure the units that they will not be punished if a pattern of gender discrimination is identified, but that institutional commitment to the ADVANCE grant means that discrepancies will be corrected to assure fair advancement of women faculty.

Simultaneously, the faculty leaders of the project need to explain the reasons for the data collection to their faculty colleagues, conveying the goodwill and positive commitment on the part of the administration to insure gender equity, using these extensive, broader measures. In discussing these issues, faculty may find compelling the ample scholarly evidence and burgeoning research (Malcom, Chubin, and Jesse 2004; George et al. 2001) from other institutions on environmental effects on productivity. The faculty leaders will need to partner with the administrative leaders to insure that faculty understand positive future intentions of the institution for senior tenure-track women.

5. Is the institution willing and able to contribute significant resources to this project?

The commitment on the part of the upper administration to identify significant funding over a long time frame provides an additional signal to the faculty of the future intentions of the institution. For example, the ADVANCE grants from various institutions propose unique ways in which the funding will be used, tailored to the particular institutional needs and plans to advance women to senior positions. Some use most of the money to establish professorships (see thread 1), while others use substantial amounts for new faculty hires and start-up packages. Most devote considerable resources to workshops, conferences, and other faculty development opportunities (thread 4). All agree to institutionalization of the initiatives begun under ADVANCE when NSF funding ends.

Agreement by the university to continue tenure-track lines and endowed professorships begun under ADVANCE constitutes a real dollar commitment. Similarly, continuing workshops and other faculty development programs, as well as course buyouts under active service–modified duties programs begun under ADVANCE, will cost the institution substantial amounts in future years. In the initial consideration and planning for ADVANCE grant submissions, most individuals recognize these future costs.

Less evident may be the substantial in-kind and real dollar costs accumulated during the actual period of NSF funding. For example, a dean or provost serving as PI or co-PI on the project may request no compensation from the ADVANCE grant, yet will spend hundreds of hours on the project over the five-year period. The time spent on ADVANCE by senior administrators, with salary paid by the university, cannot be spent on other beneficial activities. In a similar fashion, many of the family-friendly practices such as active service–modified duties must be extended to men as well as women faculty (thread 3). During the period of the ADVANCE grant, some institutions have chosen to use the NSF money for course buyout for active service–modified duties for women faculty but used institutional money for men faculty. In short, even during the period of the grant, and certainly afterward, the institutional financial commitments likely exceed those of the considerable money provided by NSF. This proportion of NSF funding relative to institutional resources, salaries, and cost-sharing suggests that many institutions could undertake an ADVANCE-like institutional transformation without external funding, particularly since the institution would experience relief from the reporting, evaluation, and travel to PI meetings required by NSF.

6. If the project received external funds, where will the home for the project be when funding ends and how will the project continue?

Although final housing for the project in terms of its organization and location on campus—the program, center, or institute where the project will continue at the end of funding—seems to be a decision that might wait until the last couple of years of project funding, considering this during the initial planning phase of the project is critical for both implementation and institutionalization phases. The ideal ultimate project home will be an existing, active, well-respected unit that is appropriately positioned in the institutional structure with sufficient resources to carry out its current functions in addition to the project.

As suggested in discussion of question 3, an extant, well-functioning, and respected women's program, or other alternatives, depending upon the university's organization and administrative culture, might serve as the perfect home for something like an ADVANCE project. Articulating the final home during the planning phase and initial implementation of the project provides numerous benefits. For example, informing the community that the extant women's program will be the final home underlines the university's commitment to and strengthening of that women's program, rather than creating a new bureaucratic entity to house the project. Knowing the structure of the final home and its relationship to overall institutional structure may help determine the level of transformation sought. If transformation of the entire institution is the goal, then the final home should not be in a center or program of only one college that reports to the dean of that college. Project leadership can include respected women faculty leaders from the women's program, both to establish credibility for the project and to facilitate a smooth transition to the final home. This articulation can explicitly demonstrate how this project to advance tenure-track faculty women to senior positions links with other campus efforts to attract, retain, and support women at different levels.

7. Will the institution undertake the project, or at least a scaled-down version of it, even if external funding is not obtained?

A negative response to this question suggests that more than budget constraints may prevent an institution from pursuing this worthy project during a time of shrinking resources. As the discussion of all of the above questions suggests, and as explicitly stated previously, projects like ADVANCE require significant institutional financial resources to complement any external funding they receive, a willingness to change policies and practices, and the buy-in of the upper administration, along with grassroots faculty support. Above all, the goal of the project must mesh with the university's strategic plan and be defined as an institutional priority.

Without this priority definition, a project will be less likely to succeed. Genuine transformation, particularly at the institutional level, remains notoriously difficult and elusive. Resistance because of competing priorities, resource constraints, and misunderstanding of the desired transformation all work to undermine change. A strong commitment, as evidenced by a decision to undertake some version of the project even in the absence of external funding, by both the upper administration and grassroots faculty, increases the likelihood of project success in the happy event that external funding is awarded.

Conclusion

A negative or lukewarm response to any of these questions suggests not only that the institution may not be ready to plan such a project, but most definitely that it has not thought through the steps needed to insure sustainability and institutionalization of the project when the NSF funding ends. Institutions seeking to replicate successful ADVANCE projects from other campuses will find that negative answers to any of those same questions indicate that they may be unable successfully to institutionalize the project they wish to replicate.

For those universities that have experienced the privilege of a NSF ADVANCE grant, the ultimate question to raise about institutionalization is whether in ten or twenty years this institution will or should remember ADVANCE. With ADVANCE project goals and objectives successfully institutionalized, these should become so well integrated that they are seen as business as usual. Is this self-antiquation, through integration into the mainstream of the institution, the ultimate goal of ADVANCE, or will ADVANCE or its successor be needed for the foreseeable future?

NOTE

Portions of this chapter have been used in an article by these same authors in the *Journal of Technology Transfer* in its special issue devoted to ADVANCE. Rosser, Sue V., and Jean-Lou Chameau. 2006. Institutionalization, sustainability, and repeatability of ADVANCE for institutional transformation. *Journal of Technology Transfer* 31:335–44.

REFERENCES

Awareness of Decisions in Evaluating Promotion and Tenure (ADEPT). http://www.adept.gatech.edu/.

Bartlett, T. 2005. More time. *Chronicle of Higher Education* 52 (2): A16.

Boxer, M. J. 1998. *When women ask the questions: Creating women's studies in America.* Baltimore: Johns Hopkins University Press.

Braxton, J. M., W. Luckey, and P. Helland. 2002. Institutionalizing a broader view of scholarship through Boyer's four domains. In *ASHE-ERIC Higher Education Report* 29, no. 2. San Francisco: Jossey-Bass/John Wiley Periodicals.

Commission on Professionals in Science and Technology. 2000. *Professional women and minorities: A total resources data compendium.* 13th ed. Washington, DC: Commission on Professionals in Science and Technology.

Cook, S.G. 2001. Negotiating family accommodation practices on your campus. *Women in Higher Education* 10 (4): 25–26.

Eckel, P., and A. Kezar. 2003. *Taking the reins: Institutional transformation in higher education.* Westport, CT: Praeger.

Fasbach, L. 2005. Where are the men? *Chicago Tribune,* February 2.

Fausto-Sterling, A. 1992. Building two-way streets: The case of feminism and science. *National Women's Studies Association Journal* 4 (3): 336–49.

Garrad, L. 2002. Berkeley, 1969: A memoir. *Women's Studies Quarterly* 30 (3–4): 60–72.

George, Y., D. Neale, V. Van Horne, and S. Malcom. 2001. *In pursuit of a diverse science, technology, engineering, and mathematics workforce: Recommended priorities for enhanced participation by underrepresented minorities.* Washington, DC: American Association for the Advancement of Science.

Heifetz, R., and D. Laurie. 1997. The work of leadership. *Harvard Business Review* 75 (1): 124–34.

Hong, P. 2004. Gender gap growing on college campuses. *Los Angeles Times,* December 1.

Malcom, S., D. Chubin, and J. Jesse. 2004. *Standing our ground: A guidebook for STEM educators in the post-Michigan era.* Washington, DC: AAAS and NACME.

Mason, M. A., and M. Goulden. 2004. Do babies matter (Part II). Closing the baby gap. *Academe,* November–December.

Massachusetts Institute of Technology. 1999. A study on the status of women faculty in science at MIT. *MIT Faculty Newsletter* 9 (4). http://web/mit/edu/fn1/women//women.html.

Merton, P, J. Froyd, M. C. Clark, and J. Richardson. 2004. Challenging the Norm in Engineering Education: Understanding Organizational Culture and Curricular Change. *Proceedings, ASEE Annual Conference.* Paper published in "Proceedings of the 2004 American Society of Engineering Education Annual Conference" held in Salt Lake City, UT.

Meyerson, D. E. 2003. *Tempered radicals: How everyday leaders inspire change at work.* Boston: Harvard Business School Press.

Pendred, D. 2004. Plight of Black Males Explored. *Atlanta Journal Constitution,* November 7, El, E7.

Pope, J. 2005. Harvard to commit $50M to women's programs. *Boston Globe,* May 17. www.boston.com/news/education/higher/articles/2005/05/17html.

Rogers, E. 2003. *Diffusion of innovation.* 5th ed. New York: Free Press.

Rosser, S. V. 1988. The impact of feminism on AAAS meetings: From nonexistent to negligible. In *Feminism within the science and health care professions: Overcoming resistance,* ed. Sue Rosser, 105–16. Elmsford, NY: Pergamon Press.

———. 2000. Editorial on women and science. *Women's Studies Quarterly* 23 (1–2): 6–11.

———. 2004. *The science glass ceiling: Academic women scientists and the struggle to succeed.* New York: Routledge.

Sonnert, G., and G. Holton. 1995. *Who succeeds in science? The gender dimension.* New Brunswick, NJ: Rutgers University Press.

Strang, D., and S. A. Soule. 1998. Diffusion in organizations and social movements: From hybrid corn to poison pills. *Annual Review of Sociology* 24:265–90.

U.S. Department of Education. 2004. The chronicle almanac: The nation. *Chronicle of Higher Education* 51 (1): 4.

Woodbury, S., and J. Gess-Newsome. 2002. Overcoming the paradox of change without difference: A model of change in the arena of fundamental school reform. *Educational Policy* 16 (5): 763–82.

Xie, Y., and K. Shauman. 2003. *Women in science: Career processes and outcomes.* Cambridge: Harvard University Press.

Measuring Outcomes

INTERMEDIATE INDICATORS OF
INSTITUTIONAL TRANSFORMATION

Lisa M. Frehill, Cecily Jeser-Cannavale, & Janet E. Malley

HOW CAN administrators know whether new strategies of faculty recruitment and retention "work" in increasing the representation of women among faculty? In this chapter we discuss ways that administrators can assess change within their institutions over time and relative to other institutions. To illustrate these procedures, we use data from nine institutions that were funded by the National Science Foundation ADVANCE Institutional Transformation program; grantees are required to annually report to the NSF data related to a set of twelve indicators.

The indicators (enumerated in appendix 1) were developed during an April 2002 meeting of the nine first-round ADVANCE awardees.[1] The MIT report formed the basis for the discussion about this list of indicators, which included information about resource allocation (space, salaries, start-up packages), rank and tenure status of women faculty, women's access to administrative positions, and so on. Because it reflects the input from the nine different—and diverse—institutions (see table 1), this list is broad and provides a basis for other institutions interested in similar assessments. Collecting these data represented a major undertaking for most institutions; however, the NSF requirements for data reporting helped ensure that institutional support was available to facilitate their collection.

There were several purposes for collecting these data, including assessment of program effectiveness and as a potential data source for understanding institutional change and gender equity. We will focus on a few of the reported indicators collected by the nine first-round institutions, for which the most complete data are available (specifically, number of faculty by rank, gender, and department)[2] as examples of how an administrator might use these data. Furthermore, we will show how these data compare to national statistics and provide examples of presentation formats used by ADVANCE awardees.

Initial work was done by Frehill and Jeser-Cannavale in collaboration with other ADVANCE project team members to assemble data for cross-institutional comparisons. The question of measurement was central in several conference papers (Frehill and Jeser-Cannavale 2004a, 2004b). After having collected and reported data for three years, they had an opportunity to consider the larger context in which these data played a role in institutional transformation.[3] It became clear that standardized measurement and reporting were essential to answer key questions about women's status relative to men:

1. To what extent are women in positions similar to those of men?
2. To what extent are the institutional processes of advancement equitable for men and women?
3. To what extent are women represented among the key decision-makers at the institution?
4. To what extent does gender affect the distribution of scarce institutional resources?

In this chapter, we address only the first two of these questions. Beyond examination of the individual ADVANCE projects, we wanted to be able to gauge the impact of the NSF ADVANCE program overall by comparing these data to those from nonawardee institutions. Despite some limitations in the available data, we will describe several comparisons as exemplars.

National Comparisons: Methodological Considerations

Unit of Analysis, Level of Aggregation, and Institutional Structures

For reasons of confidentiality, ADVANCE data (e.g., salary, tenure and promotion outcomes, and start-up packages) are mostly reported at the department, rather than the individual, level. Within the ADVANCE institutions it makes sense to use departments and colleges (or schools) as the unit of analysis. Each academic unit within the larger university can monitor its own progress toward programmatic goals, as compared to other units at the same institution. Also, in many cases, unit-level administrators (e.g., the school or college dean) have access to various mechanisms by which to implement change within their units. Deans, therefore, are often key decision-makers who are interested in gauging the "diversity performance" of disciplines within their span of control.

Data available at the national level, however, are aggregated at the disciplinary level (e.g., Engineering; Mathematics and Computer Sciences; etc.). To enable national-level comparisons, in the absence of individual-level data but given department-level data, we have aggregated data reported by the nine first-round ADVANCE awardees using these same disciplinary categories. This allows for comparisons among the NSF ADVANCE institutions as well as with national published reports of postsecondary data (e.g., Nelson 2005; NSF 2004a, 2004b) regarding women's representation on STEM faculties.

Such aggregation comes with some cautions. As demonstrated in much research on occupational gender segregation (e.g., Reskin 1984; Reskin and Roos 1990), aggregation across many categories can obscure important differences in the levels of women's representation within subcategories. Likewise, aggregation at the discipline level means that factors associated with how disciplines are organized within a university are ignored. For example, a biologist might be located in a department within an agricultural college at one institution (e.g., University of Wisconsin), or in one that is part of a school of biological sciences at another (e.g., University of California at Irvine), while at a third, she might be in a still larger unit such as a college of arts and sciences (e.g., New Mexico State University). Administrators should take into account how the different institutional arrangements might affect faculty members' careers.

NSF publications differ in aggregation level (appendix 2). The Survey of Earned Doctorates reports on detailed categories including sub-disciplinary fields, often used to establish "availability pools" for affirmative action (e.g., http://www.eod.uci.edu/availstats.html). These data are also useful for departments in strategizing searches to tap into the areas in which more women (or underrepresented minorities) are located within a particular field. But in the more comprehensive

Women, Minorities and Persons with Disabilities in Science and Engineering (NSF 2004b), broader categories are used to report information about faculty, which we use in this chapter:[4]

engineering

biological and agricultural sciences

physical, earth, atmospheric, and ocean sciences[5]

mathematics and computer science

psychology

social sciences

To some extent, these categories aggregate fields that have similar levels of women faculty. In many cases, discipline-based professional associations provide information about the representation of women and minorities for more detailed specialties within the discipline and should be consulted, for example, by chairs of departments and search committees. Deans can hold department-level administrators accountable by requesting that information from disciplinary associations be provided prior to approving faculty searches.

Time Lag in Data Availability

National-level data are often several years old by the time they are published. For example, faculty data in the most recent NSF publications report for the year 2001; data on earned doctorates are slightly more recent (2003).[6] Thus, for the first nine ADVANCE programs, begun between September 2001 and May 2002, these NSF data can only provide a limited glimpse of programmatic progress relative to any larger national changes that occurred at institutions that did not have the NSF-funded ADVANCE program.

Acquiring up-to-date data within one's own institution may also be difficult; in addition, time must be spent verifying and cleaning the data. Colleges and universities without strong institutional research units may want to consider ways to develop processes for internal data-collection and assessment. It is unlikely that sufficient data will be available to assess real progress at the institutions awarded ADVANCE grants until several years after the end of the program. Hence, results reported here are considered preliminary.

Methods for National Comparisons

For this chapter we focused on institutionally reported data for the number of faculty by sex and rank in the STEM fields by department after all departments were assigned a numeric code based on the NSF categories. Again, as mentioned above, departments were assigned to discipline categories regardless of their unit location within each institution. Not all of the NSF ADVANCE institutions targeted psychology and social sciences; therefore they may have not reported data for faculty in these fields. In addition, we omitted all departments that were housed in medical schools.

These are cross-sectional data; as such, they represent net changes in the number and proportion of women faculty rather than revealing institution's relative success or failure in terms of hiring and promotion processes (we do not report on these numbers). This is particularly important in the case of assistant professors, as net decreases in their number could be due to promotions rather than to attrition.

In our analyses, we make comparisons among the nine first-round ADVANCE awardees with reference to national average representations of women in tenured and tenure-track faculty positions in each of the broad disciplinary areas as reported by NSF (2004b) for 2001. Percentages within five percentage points of national levels were considered consistent with them, those more than five percentage points less were considered below the national average, and those at least five percentage points higher were considered above average.

Table 1 presents basic information about the first nine NSF-funded ADVANCE institutions, ordered according to the widely varying levels of 2003 academic research and development expenditures. This ordering is replicated in our subsequent results tables. As shown in table 1, the ADVANCE institutions are geographically dispersed, and all but two are Carnegie Doctoral/Research-University-Extensive institutions: UPR–Humacao is a Baccalaureate College I and Hunter is a Masters Colleges and Universities I. Both of these institutions have over 70 percent female students, and neither spent over $100 million on academic research. Georgia Tech, predominantly an engineering school, has proportionately the fewest female students and faculty compared to the other institutions.

TABLE 1. General Institutional Characteristics of "First Round" ADVANCE Awardees, Ordered by Research Expenditures

	Carnegie Classification	Number of Students		Graduate Students		Faculty		2003 Research Expenditures[a]
		Total	% Female	Total	% Female	Total	% Female	
University of Puerto Rico-Humacao	Baccalaureate Colleges, General	4,500	71	0	0	N/A	N/A	N/A
CUNY, Hunter College	Masters Colleges & Universities, I	20,679	72	5,113	78	608	52	$28.6 million
New Mexico State University	D/R–U–E	16,174	55	3,021	52	693	37	$108 million
University of California, Irvine[b]	D/R–U–E	23,290	49	3,209	47	915	24	$229 million
Georgia Tech	D/R–U–E	11,457	28	5,022	24	848	16	$346 million
University of Colorado–Boulder	D/R–U–E	29,151	47	4,611	47	1,011	27	$370 million
University of Washington[b]	D/R–U–E	39,136	52	11,836	52	3,300	30	$602 million
University of Wisconsin[b]	D/R–U–E	41,588	52	8,924	48	2,060	25	$677 million
University of Michigan[b]	D/R–U–E	38,000	51	11,147	44	2,717	26	$770 million

Note: D/R–U–E = Doctoral/Research University-Extensive. Data not available for University of Puerto Rico–Humacao on female faculty or research expenditures.

[a]Total academic research and development expenditures; source: NSF WebCASPAR.

[b]Includes a medical school.

National Comparison Results

Table 2 reports women's representation on STEM faculties at the nine first-round ADVANCE institutions in fall 2001 and fall 2004. The data are presented according to NSF discipline groupings with comparable faculty data for 2001 (NSF 2004b) and allow us to examine how the institutions compare to one another and with national averages along these dimensions.

Examining the situation for women at each of the nine ADVANCE institutions, we find broad differences in women's representation on STEM faculties. Women are much more highly represented on the STEM faculties at UPR–Humacao and at Hunter than at the other seven institutions, and, indeed, their representation on the faculties at these two schools is higher than the national figure shown in the bottom row of table 2. Perhaps because these institutions already were doing well in terms of women's representation in STEM, neither posted a substantial net gain or loss in women across the six STEM fields from fall 2001 to fall 2004. That New Mexico State was also already doing well in the social sciences—an area not targeted by NMSU's ADVANCE program—is likely due to the relative position of these disciplines within the university. This raises a critical issue when making comparisons across institutions. While these nine institutions share some characteristics with other individual schools beyond their common status as ADVANCE awardees, they differ from each other in significant ways.[7] In their own assessment of their institution's place among peers, administrators will want to select appropriate institutions for making such comparisons. Indeed, while we were working on this chapter, the Carnegie classifications were changing. In recognition of the complexity of higher education institutions, a new five-dimensional scheme by which institutions could make comparisons to different sets of peers for different purposes was developed by the Carnegie Foundation (Carnegie Foundation for the Advancement of Teaching 2005).

There were no major changes in women's proportionate representation in the physical, earth, atmospheric, and ocean sciences between 2001 and 2004 at any of the nine schools. Georgia Tech, Wisconsin, and Michigan all reported a net increase of three women in these areas between 2001 and 2004 (both Georgia Tech and Michigan moved from below the national percentage to consistent with it between 2001 and 2004): NMSU and Colorado reported slight decreases in the number of women. By 2004, women's representation was within five percentage

TABLE 2. **Women's Representation on STEM Faculties at ADVANCE Institutions, Fall 2001 and Fall 2004, by discipline**

All Tenured & Tenure Track Professors		Physical, Earth, Atmospheric, and Ocean Sciences			Mathematics and Computer Sciences			Engineering			Biological and Agricultural Sciences			Psychology			Social Sciences		
		Total	Number Female	% Female	Total	Number Female	% Female	Total	Number Female	% Female	Total	Number Female	% Female	Total	Number Female	% Female	Total	Number Female	% Female
University of Puerto Rico, Humacao	2001	36	11	30.6	19	6	31.6	N/A	N/A	N/A	33	17	51.5%	N/A	N/A	N/A	17	9	52.9
	2004	35	10	28.6	18	7	38.9	N/A	N/A	N/A	35	17	48.6	N/A	N/A	N/A	15	7	46.7
CUNY, Hunter College	2001	29	6	20.7	29	8	27.6	N/A	N/A	N/A	19	7	36.8	25	12	48.0	74	27	36.5
	2004	29	7	24.1	29	8	27.6	N/A	N/A	N/A	20	7	35.0	27	15	55.6	82	31	37.8
New Mexico State U.	2001	45	5	11.1	35	6	17.1	77	6	7.8	68	13	19.1	13	6	46.2	47	23	48.9
	2004	47	4	8.5	40	12	30.0	75	7	9.3	70	15	21.4	11	4	36.4	47	23	48.9
U. of California, Irvine	2001	82	10	12.2	68	8	11.8	82	9	11.0	85	16	18.8	39	16	41.0	134	41	30.6
	2004	88	10	11.4	87	17	19.5	89	9	10.1	98	25	25.5	43	21	48.8	197	55	27.9
Georgia Tech	2001	78	6	7.7	100	11	11.0	339	36	10.6	23	3	13.0	15	4	26.7	57	15	26.3
	2004	82	9	11.0	106	11	10.4	401	42	10.5	34	5	14.7	19	5	26.3	66	19	28.8
U. of Colorado, Boulder	2001	141	22	15.6	66	7	10.6	139	16	11.5	76	17	22.4	48	12	25.0	139	37	26.6
	2004	145	20	13.8	64	10	15.6	126	11	8.7	69	17	24.6	52	18	34.6	152	51	33.6
U. of Washington	2001	142	14	9.9	56	5	8.9	191	23	12.0	38	11	28.9	N/A	N/A	N/A	N/A	N/A	N/A
	2004	140	14	10.0	62	7	11.3	196	28	14.3	36	11	30.6	N/A	N/A	N/A	N/A	N/A	N/A
U. of Wisconsin	2001	134	13.5	10.1	87.67	6.75	7.7	192.25	13.75	7.2	343.81	59.75	17.4	35	13	37.1	238.32	54.2	22.7
	2004	136.25	16.5	12.1	86.5	7.25	8.4	199.6	24.75	12.4	392.2	66.5	17.0	36	13	36.1	239.22	61.2	25.6
U. of Michigan	2001	144	14	9.7	58	5	8.6	274	29	10.6	160	37	23.1	N/A	N/A	N/A	N/A	N/A	N/A
	2004	153	17	11.1	62	7	11.3	302	31	10.3	163	42	25.8	N/A	N/A	N/A	N/A	N/A	N/A
ADVANCE, Total	2001	831	101.5	12.2	518.67	62.75	12.1	1,294	133	10.3	845.81	180.75	21.4	175	63	36.0	706	163	23.1
	2004	855.25	107.5	12.6	554.5	86.25	15.6	1,389	153	11.0	917.2	205.5	22.4	188	76	40.4	798	247	31.0
Tenured and Tenure Track Faculty, All Ranks[a]	2001	23,800	3,610	15.2	18,520	3,040	16.4	18,340	1,290	7.0	45,830	11,370	24.8	17,940	7,770	43.3	30,580	8,510	27.8

Note: Hunter College and U. of Puerto Rico, Humacao do not have engineering programs. Psychology was targeted at Hunter, UC Irvine, Georgia Tech, and Colorado. Social sciences were targeted at UPR–Humacao, Hunter, UC Irvine, and Georgia Tech.

points of the national level for 2001 (15.2%) for all institutions except at NMSU (8.5%) and Washington (10.0%).

There were larger gains at several schools in mathematics and computer science. UC Irvine and NMSU posted substantial gains in these areas: in the first three years of the ADVANCE program, the number of women in this area at NMSU doubled from six to twelve, so that the 30% female representation in 2004 was nearly twice the national level in 2001 (16.4%). UC Irvine reported a net addition of nine women to mathematics and computer science, and Colorado had a small increase, which brought that institution's representation of women close to the 2001 national level. Georgia Tech posted no change, while at Washington, Michigan, and Wisconsin women continued to be underrepresented in comparison to the 2001 national level.

Engineering reflected the most pressing "pipeline" problem. Although women engineers are more likely than their male counterparts to pursue an academic (i.e., nonindustry) position after completing their Ph.D. (NSF 2004a), the relative numbers of women are small. As shown in table 3, women account for only 17.1% of all doctorates in engineering, the lowest of any of the NSF-recognized STEM fields. Interestingly, all seven of the ADVANCE institutions with engineering programs were at or above the 7.0% national level of women in tenured and tenure-track engineering faculty positions in 2001. Most ADVANCE institutions posted modest net gains of women faculty; only Colorado had a net loss of female engineering faculty members. The most substantial gains were at Washington and Wisconsin, where by 2004 women's representation among engineering faculty exceeded the 2001 national level by five or more percentage points.

Net gains in the number of women were posted by many ADVANCE institutions in the biological and agricultural sciences. This is, perhaps, the area in which the ADVANCE schools should have most expected to see improvement because of the larger pool of women with degrees in a broad range of life science disciplines. Women were awarded 37.5% of doctoral degrees in the biological and agricultural sciences in 1994—indicating that a substantial pool existed for some time prior to the ADVANCE award (NSF 2004a). By 2004, gains at UC Irvine and NMSU brought these previously-lagging institutions to parity with the 2001 national level of women's participation in life sciences (24.8%). Colorado, Michigan, and Washington remained on par in 2004 with the 2001 national representation; only Georgia Tech and Wisconsin continued to lag below the national level of women's representation.

Psychology and the social sciences were targeted by only a few of the ADVANCE institutions: Hunter, UC Irvine, Colorado and Georgia Tech for psychology and UPR-Humacao, Hunter, UC Irvine, and Georgia Tech for the social sciences. There were few changes in women's representation in these areas at ADVANCE schools. It is not clear why this is so. Perhaps the difference is one simply of scale. The natural and physical science disciplines may have benefited from attention to the problem as well as efforts to improve the offer and hire rates of women faculty that have become broad—even national—in scope, while similar concerns and efforts in the social sciences and psychology are more local and idiosyncratic.

Table 3 limits the cross sectional analysis to examine the number of assistant professors at each school in fall 2003 in comparison to the percentage of women among all Ph.D. recipients in those disciplines from the 2003 Survey of Earned Doctorates. The Survey of Earned Doctorates provides a crude indicator of "pool availability" of women in STEM fields. By examining the relative proportion of women among assistant professors, we can gain insight into the past several years' worth of hiring decisions at each institution, including several years prior to the ADVANCE grant. Again, administrators are cautioned to take into consideration the varied post-doctoral expectations across disciplines when looking at this publication for availability data. In many "bench sciences" (e.g., life and physical sciences) assistant professor candidates at research universities are expected to have completed one to three post-docs prior to entering the professoriate. In other fields like mathematics and computer science, postdocs are still rare.

Interestingly, women's representation among engineering assistant professors at ADVANCE institutions is almost universally above or close to national figures. Women accounted for 17.1% of engineering doctorates awarded in 2003 nationwide, their representation as assistant professors at the seven first-round ADVANCE institutions with engineering programs in 2003 ranged from about 15% (Colorado, UC Irvine, and NMSU) to 25% at Washington and Georgia Tech. This is in contrast to the recent hiring patterns of women in psychology, the STEM field with the largest pool of women (68.1% of doctorates in psychology in 2003 were awarded to women). None of the ADVANCE institutions who tracked this field matched the availability pool.

Women's representation among assistant professors in the physical, earth, atmospheric and ocean sciences varied widely at the nine ADVANCE institutions from a low of 9.1% at Georgia Tech to a high

TABLE 3. Women's Representation among Assistant Professors in STEM at ADVANCE Institutions, 2003, by Discipline

Assistant Professors 2003	Physical, Earth, Atmospheric, and Ocean Sciences			Mathematics and Computer Sciences			Engineering			Biological and Agricultural Sciences			Psychology			Social Sciences		
	Total	Number Female	% Female	Total	Number Female	% Female	Total	Number Female	% Female	Total	Number Female	% Female	Total	Number Female	% Female	Total	Number Female	% Female
University of Puerto Rico, Humacao	6	1	16.7	6	2	33.3	N/A	N/A	N/A	5	3	60.0	N/A	N/A	N/A	6	3	50.0
CUNY, Hunter College	4	1	25.0	4	0	0.0	N/A	N/A	N/A	2	0	0.0	6	1	16.7	19	11	57.4
New Mexico State U.	15	2	13.3	13	5	38.5	20	3	15.0	26	9	34.6	5	2	40.0	11	6	54.5
U. of California, Irvine	20	2	10.0	20	5	25.0	20	3	15.0	25	12	48.0	5	2	40.0	39	12	30.8
Georgia Tech	22	2	9.1	30	2	6.7	74	19	25.7	12	3	25.0	4	0	0.0	19	5	26.3
U. of Colorado, Boulder	28	2	7.1	17	3	17.6	27	4	14.8	18	8	44.4	10	5	50.0	46	26	56.5
U. of Washington	17	5	29.4	7	2	28.6	36	9	25.0	5	3	60.0						
U. of Wisconsin	24	4	16.7	11	2	14.3	47	8	17.2	72	35	48.8	7	0	0.0	70.5	29	41.1
U. of Michigan	28	6	21.4	7	2	28.6	51	8	15.7	31	11	35.5	N/A	N/A	N/A	N/A	N/A	N/A
ADVANCE, Total	164	25	15.3	115	23	19.3	275	54	21.4	196	84	42.9	37	10	27.0	210.5	92	43.7
Survey of Earned Doctorates, 2003[a]	4,092	1,151	28.1	1,857	438	23.6	5,242	896	17.1	6,600	2,911	44.1	3,271	2,229	68.1	4,126	1,847	44.8

Note: Hunter College and U. of Puerto Rico, Humacao do not have engineering programs. Psychology was targeted at Hunter, UC Irvine, Georgia Tech, and Colorado. Social sciences were targeted at UPR–Humacao, Hunter, UC Irvine, and Georgia Tech.

[a] *Source:* NSF 2004a.

of 29.4% at Washington. Only two institutions—Washington and Hunter—had representation of women among assistant professors consistent with the 28.1% 2003 availability in these fields.

Recent hiring patterns in mathematics and computer sciences and in the biological and agricultural sciences at ADVANCE institutions reflect evidence of potential progress. Women's representation as assistant professors at five of the institutions—UC Irvine, Washington, NMSU, Michigan, and UPR–Humacao—were at or above the 23.6% "availability" for mathematics and computer sciences. The same was true for UC Irvine, Colorado, UPR–Humacao, Washington, and Wisconsin with respect to the biological and agricultural sciences (44.1% of national pool).

In 2003 women earned 44.8% of social science doctoral degrees (NSF 2004a). Women's representation among social science assistant professors at five of the seven institutions from which these data were available reveal that hiring over the past several years was at or above par with the availability pools. The exceptions were Georgia Tech, where women accounted for only 26.5% of assistant professors in the social sciences, and UC Irvine, where 30.8% of social science assistant professors were women in 2003.

Peer Comparisons

As discussed earlier, and shown in table 1, the nine ADVANCE awardee institutions are connected merely by their participation in the NSF-funded gender equity program. An institution's own peer institutions—usually determined by a board of regents or institutional research office—offer a more fruitful comparison group to understand the extent of change relative to gender equity goals. In many cases, institutions may have different sets of peers for different purposes. For example, Washington has two such lists: one associated with the Office of Financial Management (a group of eight institutions) and a lengthier list associated with the Higher Education Coordinating Board (twenty-four institutions). Institutions that are part of larger systems may also find that other institutions within their systems are important units for comparison.

Such is the case with the UC Irvine.[8] As part of a large system that includes ten campuses and three national laboratories, the University of California Office of the President (UCOP) coordinates collection and reporting of institutional data, including faculty data across campuses. We accessed UC system-wide hiring data online for the four most recent academic years: 2000–2001 through 2003–4 to prepare table 4 and figure 1. Using the UCOP definitions of discipline groups—which differ from those we have used in the earlier part of this chapter—we computed UC Irvine hiring by gender for these same four years using department-level hiring data provided in the UC Irvine ADVANCE annual reports to NSF.

In this case, the key research question of interest is this: *To what extent did UC Irvine, as the only ADVANCE awardee in the UC system, hire women at a higher rate than the other UC campuses?*

Even though the mix of disciplines varies across the campuses, at least a number of other institutional factors are "controlled" in this analysis by virtue of the central organization of the system. In addition, a recent report (West et al. 2005) on gender discrimination in hiring in the UC system made national headlines (Jaschik 2005; Lewin 2005). It is interesting to note that many remedies suggested by West et al. (2005) to rectify gender discrimination in hiring had been implemented by the UC Irvine ADVANCE program. School-based "equity advisors" are the heart of the UC Irvine ADVANCE program. As a "faculty assistant to the dean," an equity advisor works with search committees to implement various best practices in hiring. They are both a resource and an important search oversight mechanism that provides an added voice for gender equity to be heard within the search process.[9] Hence, this analysis can shed light on whether such proposed strategies enacted at UC Irvine were effective, as similar programs were not initiated at other UC campuses.

TABLE 4. **Women as a Percentage of Regular Rank Faculty New Hires, 2000-2002 and 2002-2004, and Change over Time, UC Irvine and University of California, System-Wide**

	UCI		UC System-Wide		Change (2002–04 to 2000–02)	
	2000–02	2002–04	2000–02	2002–04	UCI	UC System
Arts & Humanities	47.5	54.8	48.6	47.4	7.3	−1.2
Social Sciences & Psychology	50.0	27.6	38.4	43.3	−22.4	5.0
Mathematics	0.0	50.0	2.8	20.0	50.0	17.2
Engineering & Computer Science	10.0	22.2	12.0	19.1	12.2	7.1
Life Sciences	14.3	61.1	25.0	42.5	46.8	17.5
Physical Sciences	0.0	8.3	16.8	19.0	8.3	2.2
Medicine	19.0	28.6	24.2	30.6	9.5	6.4

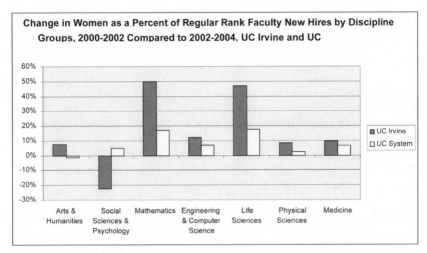

Figure 1. Comparing One ADVANCE Institution to the System-wide Change in Women's Hiring by Discipline

Table 4 shows the percentage of women among newly hired faculty at UC Irvine and across the entire UC system for two two-year periods: 2000–2002 (before the ADVANCE program at UC Irvine) and 2002–4 (the first two years of the ADVANCE program). The last two columns show the change in hiring over these two periods for UC Irvine and the UC system by subtracting the percentage of women among new hires in the 2000–2002 period from those in the later period, 2002–4. The graph in figure 1, then, illustrates the changes reported in the last two columns of table 4.

Figure 1 shows that, with the exception of the social sciences and psychology, there were proportionately more women among new hires at UC Irvine in the ADVANCE period than in the two-year period prior to ADVANCE and that this change surpassed that experienced by the rest of the UC system. Similar to the national-level comparisons discussed earlier, the greatest impact has been in those STEM fields in which the pools of available women are largest—that is, the life sciences and mathematics—but moderate increases in the rates of women's hiring are also evident in engineering and computer science, physical sciences, and medicine. While women's representation among new hires in the arts and humanities declined slightly across the UC system, there was a slight increase at UC Irvine.

Because we have located an appropriate set of peers, the analysis shown in table 4 and figure 1 represents stronger evidence for the posi-

tive impact of the ADVANCE program on gender equity in recruitment. While this is good news from a programmatic standpoint, it is still
important to recognize that hiring is merely one of the many issues
about which we were concerned. There were four questions that we
suggested should frame an analysis of the status of women at an institution. We now turn our attention to how one might examine changes
within one's own institution over time to document positive change in
the status of women faculty.

Internal Assessments

Making comparisons to peer institutions and national averages is a valuable exercise for administrators to assess how their home institutions are
progressing relative to these external yardsticks. However, as already
mentioned, important differences among academic institutions and the
limits of the data available render these comparisons less than perfect.
Looking at change within the institution is another useful strategy for
administrators.

NSF ADVANCE projects generally designated the year prior to ini-
tiation of the grant as the baseline year, and each subsequent year was
compared to this baseline. Year-to-year assessments are unlikely to
demonstrate significant change; however, by the fourth and fifth years
of their projects, NSF ADVANCE institutions were able to begin to
identify instances where real progress had been made, as previously
noted (and other instances where change was still not evident). Having
data at the department (as opposed to school- or university-wide) level
allows chairs and deans to identify where interventions may be most
helpful. In addition to total numbers of faculty on the roster, administrators may look within ranks (including time in rank at the associate
professor level) and at new hires to help determine where problems may
lie in the promotion and tenure processes.

Sex Ratios. Beyond raw and proportional data, it is useful to consider
ways to frame or consolidate the data to make them more easily com-
prehensible and meaningful. Frehill and Jeser-Cannavale (2004a), drawing on the work of Kanter (1977), recommend sex ratio categories, useful for assessing critical mass (generally considered at about 30%). For
example, departments can be categorized as those with women in pro-
portions below critical mass (i.e., 0–17%, which Kanter refers to as
"female token"); those approximating critical mass (i.e., 18%–35%,
which Kanter refers to as "female minority") and those closer to a sex-

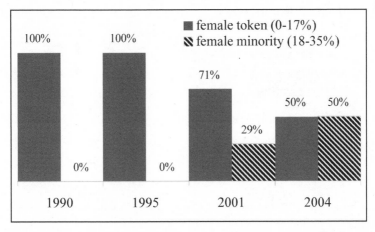

Figure 2. Percentage of Departments by Sex Ratio Categories in One ADVANCE Institution over Time

balanced ratio (36–64%). Figure 2 demonstrates the change in sex ratio pattern for the Division of Natural Science departments within the College of Literature, Science and the Arts (LSA) at the University of Michigan.

In this example the college benefited from the availability of institutional data prior to the start of the ADVANCE project, permitting a longer-term view of change. The data indicate that in the approximately ten years prior to the ADVANCE grant women were significantly underrepresented on the faculty in all natural science departments in this school. Between 1995 and the start of the ADVANCE initiative, some positive change was evident with approximately one-third of the departments moving from the female token category to the female minority category (critical mass). In the four years of the ADVANCE grant even more progress was documented, and by 2004 half of the natural science departments were in the female minority category, in the neighborhood of critical mass.

It is useful to remember that the designation of categorical percentages is based on the notion of critical mass and assumes that a gender-balanced representation is a reasonable and achievable goal. Other targets may be more appropriate for more refined assessments. For example, goals could be based on the percentage of women in the available pool of graduate students or postdoctoral fellows in a particular discipline.

Cohort Analyses. Tracking a cohort of faculty with longitudinal data

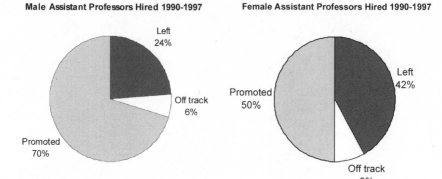

Figure 3. Proportion of Faculty Cohort Retention Outcomes by Gender, 2003 Academic Year.

is another useful approach. For example, administrators can assess retention and promotion outcomes by gender for a faculty cohort hired as assistant professors within a particular time period (e.g., five years) and assess their status after a time period sufficiently long to include tenure review. Figure 3 shows such an analysis for faculty hired as assistant professors between 1990 and 1997 for one ADVANCE college. These data reveal that by the 2003 academic year, women were promoted less often than men (50% vs. 70%) and proportionately more women both left the university and left the tenure track.[10] Again, these analyses benefit from the availability of institutional data collected prior to ADVANCE. and point to the value of instituting procedures for systematically collecting institutional data.

Flow Charts. Another example of assessing change in faculty over time is shown in the figure 4 flow chart[11] that represents tenured and tenure-track ("tenure eligible") as separate streams of movement up or out of the tenure-track system. This chart represents the change in the University of Michigan's LSA Natural Sciences Division faculty during academic year 2003–4 and shows women's movement into, up, or out of the tenure track. Within each stream we determined the representation of women in each of the four possible transition actions: hiring, tenure denial, voluntary leaving,[12] and promotion. It is interesting to note that this chart differentiates between those who voluntarily left the university and those who were denied tenure. In this case, three new tenured women were hired into LSA, one was tenured from among those who were tenure-eligible, and none of the tenured women pro-

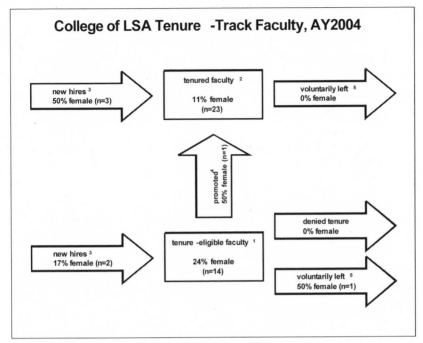

College of LSA Tenure -Track Faculty, AY2004

new hires [3]
50% female (n=3)

tenured faculty [2]

11% female
(n=23)

voluntarily left [5]
0% female

promoted[4]
50% female (n=1)

denied tenure
0% female

new hires [3]
17% female (n=2)

tenure -eligible faculty [1]

24% female
(n=14)

voluntarily left [5]
50% female (n=1)

[1] faculty at the rank of assistant professor as of AY2004;
[2] faculty at the rank of associate and full professor as of AY2004;
 faculty with 0% FTE are excluded;
 faculty with joint appointments are counted in both departments.
[3] faculty hired at the rank of assistant professor effective AY2004 or later.
[4] faculty promoted from assistant to associate professor effective AY2004.
[5] faculty terminated effective AY2004.

Figure 4. Flow Chart of One ADVANCE College

fessors left LSA between 2003 and 2004. The flow chart, then, provides a picture of women's retention and advancement, which are two of the other important goals of the ADVANCE program and are part of the answer to the second question we posed at the start of the chapter.

In this example only one academic year is represented, but charts could be constructed that reflect longer time periods (e.g., five years). The relative size of the faculty at a particular institution and the timing of programmatic efforts to address gender inequity are important considerations in determining the time frame. For example, with a smaller faculty, one may need to aggregate multiple years to preserve confidentiality and to have meaningful comparisons.

Conclusions

Although the data presented here are preliminary, the analyses are useful examples of how one might compare institutional to national data. Some conclusions about the impact of ADVANCE can be drawn from these analyses. First, for those institutions in the midst of a wave of hiring, ADVANCE has brought gender equity issues to the attention of those making hiring decisions. Not surprisingly, those areas with larger pools of women Ph.D.s—notably the life sciences—were the areas in which many ADVANCE institutions saw important gains in the hiring of women faculty.

But even in disciplines such as engineering, where a smaller proportion of women are emerging from the pipeline, creating a challenge for those who are trying to recruit women to the professoriate, ADVANCE institutions, on average, demonstrated higher percentages of women faculty than the national averages. For those ADVANCE institutions where the proportion of women as assistant professors was small (UC Irvine, Colorado, and NMSU), we found that the overall representation of women on engineering faculties was at or above the national average of 7.0% in 2001.

The findings for psychology were interesting because of the high availability of women (68.1%) in the pipeline. It was surprising that even when psychology was targeted at four ADVANCE schools, women were underrepresented among assistant professors. It could be that once "critical mass" is achieved, faculty become complacent about gender equity in hiring. Hence, the example of psychology may indicate that advocates of gender equity need also to be attentive to those fields that seem to be progressing as a result of interventions. It may also be the case that more subtle issues of gender are at play that permeate all areas of academia to which administrators should attend.

This chapter has also illustrated how comparisons to one's peer institutions can be an important process to assess change, especially if comparable data are available. For UC Irvine it is fortuitous that UCOP serves as a central warehouse for information and provides a central set of policies that guide university policies and procedures. Here we have shown that the proportion of women faculty hired has increased more at the ADVANCE awardee institution, UC Irvine, than across the UC system as a whole. Other institutions should consider accessing peer institution information via the NSF WebCaspar data warehouse (http://webcaspar.nsf.gov). Although time-consuming, obtaining peer comparison data to assess programmatic impact may be worth the effort.

Finally, we have discussed the inherent limits of national and peer comparisons and provided a range of strategies administrators might use to assess women's status across time within their own institutions. Such assessments are a means to portraying trends in hiring and promotion to audiences within our own institutions. These analyses can be conducted at the university, college, or department level and thus may be helpful for identifying specific areas where interventions may have been successful as well as areas where further efforts may be needed.

In this chapter we have only scratched the surface of understanding women's status as faculty members. Indeed, we have only addressed two of the four interrelated questions that we posed at the opening of the chapter. To gain a broader, multidimensional appreciation for women's status within their own campuses, administrators will need to attend to and make sense of a range of information that addresses questions about women's advancement and recognition and about the equitable allocation of resources. ADVANCE program personnel have grappled with these issues for several years and have conveyed guidance via the two toolkits mentioned earlier (Frehill et al. 2005). These toolkits can be useful for academic administrators interested in documenting changes in the status of women at their own institution and how that status compares to national-level trends. The specific data analyzed, and the specific comparisons made (over time, to other units within the university, to peer institutions, to national averages), matter much less than does the practice of systematically paying attention to data that indicate how well an institution is meeting its own goals.

Appendix 1

1. Number and percentage of women faculty in science or engineering by department

2. Number and percentage of women in tenure-line positions by rank and department

3. Tenure promotion outcomes by gender

4. Years in rank by gender

5. a. Time at institution by gender

   b. Attrition by gender

6. Number of women in S & E who are in non-tenure-track positions (teaching and research)

7. Number and percentage of women scientists and engineers in administrative positions

8. Number of women S & E faculty in endowed or named chairs

9. Number and percentage of women S & E faculty on promotion and tenure committees

10. Salary of S & E faculty by geinder (controlling for department, rank, years in rank)

11. Space allocation of S & E faculty by gender (with additional controls for department, etc.)

12. Start-up packages of newly hired S & E faculty by gender (with additional controls for field or department, rank, etc.)

Appendix 2

NSF AGGREGATION OF ACADEMIC FIELDS

Many National Science Foundation reports aggregate to the level that is bolded, below.

Engineering

Aeronautical/astronautical engineering

Chemical engineering

Civil engineering

Electrical engineering

Industrial engineering

Mechanical engineering

Materials/metallurgical engineering

Other engineering

Physical Sciences[13]

Astronomy

Chemistry

Physics

Other physical sciences

Earth, Atmospheric, and Ocean Sciences

Atmospheric sciences

Earth sciences

Oceanography

Other environmental sciences

Mathematical and Computer Sciences

Mathematics

Computer science

Biological and Agricultural Sciences

Agricultural sciences

Biological sciences

Psychology

Social Sciences

Economics

Political science

Sociology

Other social sciences (anthropology, area studies, criminology, geography, statistics, urban affairs/studies, social sciences, general, social sciences, other, history and philosophy of science and technology, linguistics, American studies, archaeology)

Non-S&E

Health and medical sciences

Humanities

Education

Professional/Other

Business management and administrative services

Data processing

Information fields (e.g., communications)

Other professional fields (including architecture environmental design, home economics, law, library science, parks/recreation/leisure/ fitness, social service professions, theology and religious education)

NOTES

1. The nine first-round ADVANCE institutions are University of California, Irvine; University of Colorado, Boulder; Hunter College, City University of New York; University of Michigan; New Mexico State University; University of Puerto Rico, Humacao; University of Washington; and University of Wisconsin.

2. Institutions may also want to track race or ethnicity in addition to gender. Small sample sizes are particularly acute in this case, as well illustrated by Donna Nelson's (2005) analysis of gender and ethnic compositions of "top fifty" departments in various science and engineering disciplines.

3. Efforts were made to increase standardization through a series of meetings with key project staff from several ADVANCE initiatives. These meetings led to the development of two "toolkits" (Frehill et al. 2005) to guide ADVANCE program personnel in collecting and reporting data, which are available at www.advance-portal.net.

4. While these broader categories are helpful, they may also obscure important difference among institutions. For example, because computer science is a division in a larger department in the University of Michigan's School of Engineering, computer science data cannot be included in the "mathematics and computer science" category reported here for this school. Other differences among institutions occur in different categories; thus, it is extremely important for administrators to understand how other institutions report data when making specific comparisons.

5. As noted in appendix 2, many publications do disaggregate the physical sciences from the earth, atmospheric, and ocean sciences. We found, however, that there were substantial similarities relative to the status of women in these areas, so we have aggregated them here.

6. The 2003 data were the most recent at the time of this writing. Data for 2005 have since become available.

7. As shown in table 1, Michigan, Washington, and Wisconsin are the three "closest" institutions. All are comprehensive and have medical schools, similar numbers of students, and very large expenditures for academic research and development. Indeed, the University of Washington lists Michigan and Wisconsin among its official peer institutions.

8. Hunter and UPR–Humacao are also part of systems that constitute "peers." Other institutions, like Wisconsin, Michigan, and NMSU were "flagship" campuses of larger systems in which the other campuses were not comparable. Administrators should be aware of peers and seek mechanisms of sharing information for comparison purposes.

9. It should also be noted that the UC Irvine ADVANCE initiative involves all units on campus (the university is organized into ten schools of fairly homogeneous discipline groups). Matching funds were contributed for the two schools with non-STEM fields by the executive vice chancellor at the program's inception.

10. "Off track" in figure 3 refers to the practice whereby faculty can remain at the institution, but move from a tenure-track position to another faculty track such as research or clinical.

11. The flow chart represented here is a modification of work originally done at Columbia University.

12. Retirement and other types of voluntary leaves should be disaggregated. Many ADVANCE institutions (e.g., Michigan, NMSU, Washington, and UC

Irvine) have implemented exit interview procedures to understand why faculty leave their institutions. The issue of gender differences in early retirement might also be an important area to explore.

13. In some reports, these two areas are distinct, while in other reports, the data for both are aggregated. When we examined the national-level data for these areas in comparison to the ADVANCE institutions separately, there was no meaningful variation. Therefore, in this chapter, we have combined these areas.

REFERENCES

Carnegie Foundation for the Advancement of Teaching. 2005. 2005 Carnegie classifications: Preliminary information. http://www.carnegiefoundation.org/Classification/2005-preliminary.htm. Accessed October 13, 2005.

Frehill, Lisa M., and Cecily Jeser Cannavale. 2004a. Measuring the status of women: Towards cross-institutional analysis to understand institutional transformation. Paper presented to the Annual Meetings of the Eastern Sociological Society, New York, February 21.

———. 2004b. Measuring the status of women: Towards cross-institutional analysis to understand institutional transformation. Paper presented to the ADVANCE Conference, Atlanta, April.

Frehill, Lisa M., Cecily Jeser-Cannavale, Priscilla Kehoe, Sheila Edwards Lange, Jan Malley, Ellen Meader, Jennifer Sheridan, Abigail Stewart, and Helena Sviglin. 2005. Toolkit for reporting progress toward nsf ADVANCE: Institutional transformation goals. http://www.advance-portal.net.

Jaschik, Scott. 2005. Stalled progress. *Inside Higher Ed,* May 19. http://www.insidehighered.com/news/2005/05/19/women.

Kanter, Rosabeth Moss. 1977. *Men and women of the corporation.* New York: Basic.

Lewin, Tamar. 2005. University of California faulted on hiring of women. *New York Times,* May 18, B9.

Massachusetts Institute of Technology. 1999. A study on the status of women faculty in science at MIT. *MIT Faculty Newsletter* 9 (4). http://web/mit/edu/fn1/women//women.html.

National Science Foundation, Division of Science Resources Statistics (NSF). 2004a. *Science and engineering doctorate awards: 2003.* NSF 05–300, Project Officer Joan Burrelli. Arlington, VA.

———. 2004b. *Women, minorities and persons with disabilities in science and engineering: 2004.* NSF 04–317. Arlington, VA.

Nelson, Donna J. 2005. A national analysis of diversity in science and engineering faculties at research universities. http://cheminfo.chem.ou.edu/~djn/diversity/briefings/Diversity%20Report%20Final.pdf.

Reskin, Barbara, ed. 1984. *Sex segregation in the workplace: Trends, explanations, remedies.* Washington, DC: National Academy Press.

Reskin, Barbara, and Patricia Roos, eds. 1990. *Job queues, gender queues: Explaining women's inroads into male occupations.* Philadelphia: Temple University Press.

West, Martha S., Gyöngy Laky, Kari Lokke, Kyaw Tha Paw U, and Sarah Ham. 2005. *Unprecedented urgency: Gender discrimination in faculty hiring at the University of California.* Davis, CA: Martha S. West.

NINETEEN

Maximizing Impact

LOW-COST TRANSFORMATIONS

Lee Harle

ONE OF THE goals, indeed requirements, of the ADVANCE Institutional Transformation program is to disseminate programs that have been successful in recruiting, retaining, and promoting women faculty in science and engineering to the university community. The purpose of this chapter is to highlight a number of programs that have had a high impact relative to their financial and personnel costs. For example, many programs focus on community-building through professional development and informal meetings over meals. Networks are quick to form, and new faculty experience near-immediate benefits. By encouraging institutions outside of the ADVANCE network to promote similar activities, the program hopes to contribute to a fundamental transformation of community-building within the science and engineering academic enterprise nationwide.

Programs for Faculty

New faculty often comment that on their first day, they are given the keys to their new office and told, "See you in six years at your tenure review." At many institutions, they find little or no guidance regarding the tenure and promotion process, and rumors and myths fill the vacuum. In most cases, senior faculty and administrators assume that tenure requirements and the mechanics of the promotion process are obvious, but they usually are not. ADVANCE programs at many institutions have developed a variety of low-cost activities to address this need, some of which are highlighted in this section. These include mentoring, networking, and professional development. Tenure, promotion counseling, and grant writing skills are also targeted.

In addition to programs for junior faculty, several institutions have created activities that benefit all faculty through community-building and provide venues for professional development. Many new faculty find them particularly useful for networking outside of their disciplines and improving communication and collegiality. Broad-based activities that work to influence large numbers of faculty by building a stronger community have excellent potential for achieving institutional transformation. These range from formally organized activities such as lunch series with speakers and workshops to more informal gatherings. The more formal gatherings usually require more advance preparation for topic and speaker selection. Both can have significant effects.

The activities discussed here require minimal cost and setup. A professional group such as the University of Michigan's Network of Women Scientists and Engineers, for example, can become self-sustaining once it is initiated by ADVANCE, with lasting impact and minimal cost. In some cases, faculty must be paid for their contributions, but ADVANCE typically organizes and advertises the events, and senior faculty often volunteer their expertise for mentoring, networking, and advising on promotion. If a welcome reception is organized for the fall, for example, the cost includes organizing and advertising, and providing a catered meal. Professional development workshops may also run from low to moderate cost. Materials developed by other institutions, such as those developed by Hunter College and discussed elsewhere in this book, may be adapted for use at no cost. When outside experts are brought in for the workshops, expenses can be kept reasonable by sharing them among several sponsoring units or departments.

Mentoring

Approximately eighty men and women, drawn from the STEM tenure-track faculty population of about three hundred people, participate in an ADVANCE mentoring activity at New Mexico State University. The participants complete a questionnaire with which they are assessed and assigned to an advisee. Mixers and informal lunches covering various discussion topics are organized, and many mentoring relationships have resulted. Events such as potluck dinners have been organized to bring together faculty members and their families. These are not only popular, but also low-cost, and they foster network building. For example, they helped develop a collaboration among four women engineering faculty from four different departments, three of whom were junior faculty. This support network has been extremely vital to its members' professional survival and development.

Networking

Networking can happen spontaneously whenever groups of people are brought together, but it can also be catalyzed by a formally organized event or series of events. Networking is driven by groups larger than those typical of mentoring activities and has benefits beyond mentoring. Participants benefit by meeting new individuals and gaining access to new organizations across campus. The University of Michigan and Virginia Tech have both developed venues to foster networking among women faculty, ADVANCE, and various women's groups on campus. These activities have led to the development of leadership workshops and mentoring initiatives sponsored by ADVANCE, as well as collaborations with the universities' women's centers and programs. At Michigan, the Network of Women Scientists and Engineers, composed of tenure-track women faculty in science and engineering across the entire campus, meets several times each year to talk about issues the members have in common and develop plans for the future. For example, during the first year of its existence, the network asked ADVANCE to provide leadership workshops, which were planned and paid for by ADVANCE. Network members, who were inspired and motivated by what they learned, took the initiative the following year to design an extensive weekend leadership retreat of their own, drawing on financial contributions and staff support from ADVANCE but choosing all panelists and topics themselves.

At Virginia Tech, AdvanceVT cosponsors two informal receptions, one to welcome new faculty and graduate students in the fall and one during Women's Month in the spring. A number of groups cosponsor these events, including the Women's Center, the Graduate School, the Women's Studies program, the Organization of Women Faculty, and Women in International Development. These events provide an opportunity for the women's groups to interact, introduce AdvanceVT across the campus, and introduce the new women faculty and students to the resources available to them. Involving cosponsors across campus taps into resources not available in many science and engineering departments, providing opportunities to reconnect with colleagues and meet new people.

Professional Development Workshops

Several institutions have developed workshop activities covering topics that particularly target pre- and early post-tenure faculty, such as funding and proposal writing, negotiation, mentoring, tenure and promotion advising, creating a teaching portfolio, managing a lab, advising graduate students, and communication skills. They can be tailored to target faculty's specific needs based upon feedback, and are an example of how ADVANCE can employ homegrown talent and expertise, as many institutions have found that faculty are eager and able to assist in organizing and producing the workshops. Peer-to-peer interaction often appeals to junior faculty. For example, the University of Washington ADVANCE Center for Institutional Transformation (CIC) has found that the junior faculty prefer to hear from their peers rather than an external consultant. Faculty from twenty-one different departments organize and run the workshops, and the peer-to-peer format resonates in a very meaningful way.

The University of Rhode Island ADVANCE has partnered with their Research Office to sponsor workshops covering many of these same topics. The well-received Negotiations workshop is facilitated by faculty members with training in negotiations and focuses on women, using a panel format with role play and discussion. Their Mentoring workshop is also very popular, facilitated by three experienced faculty mentors and three new mentees. Following a panel presentation, the audience participates in small-group discussions of specific mentoring scenarios. The workshop concludes with the participants generating a best-practices list for mentors.

Other institutions have utilized both internal and external workshop facilitators. For example, ADVANCE at the University of Maryland, Baltimore County, has partnered with other groups on campus such as the Faculty Development Center to cover the costs of sponsoring outside speakers, although internal faculty are used as well to facilitate their Faculty ADVANCEment workshops. The workshops have sparked follow-up discussions and encouraged greater transparency and clarity of expectations. Tenure workshops are facilitated by two senior faculty who have been advising junior faculty informally for many years, allowing them to formally disseminate their wisdom to a broader audience. In addition, the men faculty have become engaged in the ADVANCE program by attending these workshops with their women colleagues. This activity promotes the understanding that professional development activities can benefit the entire community, both women and men.

Prior to the beginning of Utah State University's ADVANCE program, training on grant tools available through the Vice President of Research (VPR) office was underutilized and largely invisible. In response to faculty requests, ADVANCE cosponsored and organized research workshops. The VPR office recognized the high level of interest and the efficiency of this program in training faculty, and hence has assumed the responsibility of organizing a research workshop week. The workshops focus on helping faculty be more efficient and effective in the grant and funding process. Topics include using research database systems, grant-writing skills, technical writing skills, working with NSF, NIH, and other funding agencies, and experiences of faculty who have been successful in receiving high-profile research awards and grant funding.

This VPR-organized forum allows senior faculty to disseminate their expertise to other faculty members and lets junior faculty make more effective use of their precious time. The VPR office has become more efficient in assisting faculty, thanks to a venue that promotes two-way communication between faculty and the VPR/Sponsored Programs Office (SPO) and presenters. High-level buy-in at the VPR level ensures the visibility and continued success of this program and encourages community-building across all levels of the faculty.

Tenure and Promotion Advising

Several institutions have instigated simple yet effective efforts to assist pre- and early post-tenure faculty in navigating the tenure and promotion process. The University of Washington, Georgia Tech, and New Mexico State University have all developed similar Professional Career Development Consulting programs to provide advice to tenure-track faculty on their tenure and promotion packages while creating transparency in the promotion process. Senior faculty who either are currently serving or have previously served on their respective college's promotion and tenure committee discuss career paths with individual faculty members. Drawing upon this unique perspective, the senior faculty provide feedback on preparing and strengthening the curriculum vitae and navigating the promotion process. At Georgia Tech, the program includes workshops on demystifying the promotion and tenure process led by professional development consultants. With buy-in from the provost and the deans at New Mexico State, deans and department chairs participate and encourage their faculty to participate. With this kind of top-down support, these events are very well received.

Utah State University developed an informal format for promotion advising after ADVANCE discovered, based on interviews and collected data, that mentoring drops off dramatically post-tenure. They discovered that getting a promotion advisory committee together was difficult, due in part to inaction by department heads. They organized a brown-bag lunch at which the vice provost and women full professors provided advice to women associate professors seeking promotion. Two women who attended formed committees and were promoted to full professor, and several others have formed and met with promotion advisory committees in preparation for being considered for promotion. The informal setting and small group size allowed women to voice their individual concerns freely, without feeling inhibited. They were able to ask very specific questions and receive feedback on what they could do to address concerns and move the process forward. The brown-bag session also allowed central administration to hear about faculty concerns. ADVANCE found that it can be very beneficial to place small groups of women or marginalized faculty in a comfortable setting in which they receive advice on faculty policies and career development directly from the university administration.

As a result of the brown-bag session and the gathering of other institutional data, ADVANCE took the initiative to make changes to the section of Faculty Code relating to promotion and tenure to provide more mentoring to associate professors. Associate professors will now form promotion advisory committees no later than three years after tenure. This low-cost event, simple to organize and advertise and held in a nonthreatening environment, has catalyzed a significant improvement in the way advice on promotion is communicated to post-tenure faculty.

Several institutions, including the University of Rhode Island and the University of Washington, have implemented highly successful, well-attended lunch series for women faculty and graduate students, primarily from the STEM community but occasionally for the entire campus. The lunches establish venues for panel discussions and presentations on a variety of topics. The University of Rhode Island "Topical Lunch Series" covers a wide array, from paper publishing to dual-career and family-friendly issues, to navigating tenure and promotion. Although one external speaker has been invited, typically internal "experts" in each area have facilitated the lunches. The ensuing discussion and professional networking have inspired faculty, particularly new STEM faculty, to make connections with other women outside their own areas. The University of Washington CIC "Mentoring for Leadership" series invites woman scholars in leadership positions to discuss their career trajectories and what they enjoy about leadership. Both internal and external speakers who are deans, presidents, or faculty who have studied leadership are invited. The speakers tell their personal stories, how and why they have done the things they have, and what they would do differently if they could. Distinguished speakers who visit campus to give seminars are asked to make time in their schedules for lunch with the women faculty. The lunch has turned into a group mentoring experience, with attendees networking and building relationships among themselves, as well as with the speakers.

Whether lunch seminars cover a wide variety of issues or focus exclusively on one, such as leadership, the impact on professional development is similar; mentoring and networking happen naturally when groups of faculty are brought together and encouraged to take advantage of the opportunities to build community.

At Utah State University, the Biology department faculty have established coffee hours, because the department's thirty-eight professors overwhelmingly desired more communication and collegiality and decided as a group that they needed a faculty lounge and social time. Weekly coffee hours are held on Friday mornings, with the faculty on each floor of the Biology building organizing the coffee hour for one month. They plan the food and prepare the coffee; each floor pools money together to pay for the food and drink. The junior faculty, especially, are regular attendees and enjoy the chance to interact with other faculty in an informal manner to ask questions and learn more about the department and university on a regular basis, without having always to initiate the interactions. The venue provides a time for faculty to relax together and network in an informal atmosphere without making anyone feel overwhelmed or obligated. Because this event is held within a single department, a sense of belonging and cohesiveness is achieved. Organizing a given month's events increases interaction among the faculty, even across research fields, on the floor that is responsible. They are highly motivated to take ownership of and continue the activity because they see immediate benefits in increased communication and community building. This is an ADVANCE-catalyzed social activity that has become self-sustaining.

Programs for Administrators

A number of institutions have successfully implemented leadership programs for department chairs and deans, including orientations for new department heads, leadership workshops, and informal monthly gatherings for women department chairs. They are often implemented top-down with buy-in and participation at the provost and dean levels, resulting in resource support and investment university-wide. Focusing on departmental challenges such as the professional development of new women faculty allows ADVANCE teams to connect with department and college leadership and convey to them the goals of ADVANCE. The participants appreciate the opportunities to interact with their peers and observe how other departments approach and resolve challenges common to all. These venues are excellent for cross-disciplinary networking.

Cost for many of these activities is incurred by providing food, which can be minimal for brief events. Most of these activities are organized and presented top-down, or very informally among the women chairs themselves, where the cost is simply a commitment from ADVANCE to get the ball rolling. Heavier personnel investment, such as that required by the workshops, is typically undertaken by upper administration personnel who organize and run the workshops.

Because academic experiences typically do not prepare faculty leaders for the issues they must confront as department heads, the New Department Head Orientation developed by AdvanceVT at Virginia Tech draws upon the knowledge of more experienced department heads. It builds a bridge between the administration and these new leaders, preparing them to fulfill their responsibilities as department chairs. Speakers at the orientation present on a variety of topics, and breakout sessions are held, focusing on personnel issues, promotion and tenure, conflict resolution, dealing with staff, and dealing with students. Gender equity issues are built into presentations on conflict resolution, effective communication, and other topics presented by different vice presidents and vice provosts. A reception with the president provides an excellent chance for those new to the university to meet the president and their peers. The orientation program is very well received and is relatively low cost.

The University of Washington has developed Quarterly Leadership Workshops for department chairs. Department chairs are responsible for organizing and presenting the workshops. Assignments rotate among the departments, allowing for a fair distribution of the workload as well as the introduction of fresh perspectives. Topics discussed at these half-day workshops include equity, leadership, professional development, group mentoring and networking, hiring, and retention. About fifty participants—deans, chairs, and emerging faculty leaders—typically attend. This is an excellent example of an activity that allows for cross-disciplinary networking and learning, as peers come together to discuss common challenges on an ongoing basis. In addition, this program has been integrated into the Training and Development Office's Strategic Leadership Program for faculty throughout campus. Handouts and presentations from the leadership workshops are available online at http://www.engr.washington.edu/advance/workshops/.

An informal yet effective activity is the monthly brown-bag lunch for women department chairs organized by the University of Michigan ADVANCE program. The women chairs meet once per month to share experiences and advice. This lunch is low-cost and easily organized, yet it provides a low-stress, informal venue for community building and allows for cross-disciplinary discussion of issues in a sustained networking environment that continues from month to month. It could serve as a model for informal gatherings for other women leaders on campus, such as deans, provosts, and emerging faculty leaders.

Institution-Wide Efforts

One of the goals of ADVANCE is to review existing university policies and practices and, where necessary, to improve them. These institution-wide efforts can aid the entire university community by addressing issues concerning recruiting, hiring, retention, and promotion and tenure processes. Identifying forms of bias, expanding recruiting efforts outside of the traditional methods, modifying language, and promoting dual-career hiring are examples of successful ADVANCE efforts. In each case, support has been provided at the provost, dean, or department administration levels. As described earlier in this chapter, this level of support increases acceptance of changing practices throughout the institution.

Costs for these activities can be kept moderate, and in some cases, these interventions require no financial contribution. For example, an interactive, web-based tool designed to help faculty and administrators recognize and reduce bias in promotion and tenure, developed at Georgia Tech and described elsewhere in this book, can be downloaded from their web site for free. A significantly more expensive recruitment program at Virginia Tech incurs costs for hosting potential faculty candidates, but it also has significant potential to improve hiring and networking both for the individuals directly involved and for their broader research community.

Simply doing a review of policy can be symbolically important; it can also be practically important and low-cost. The University of Michigan ADVANCE team has noted negative influences language may have on community discourse and has taken steps to introduce more constructive phrasing. For example, switching from "two body problem" to "dual career program" and from "trailing spouse" to "partner" can have positive effects. Describing "time off the tenure clock" as "compensation for time lost to work" changes it from *extra* time to *compensatory* time. Equally, changing "spouse" to "partner" makes family-friendly policies more inclusive of unmarried partners. These simple shifts in language can change the way we think about these issues, thereby affecting community attitudes and contributing to institutional change at a fundamental level.

Further examples of practices that have been implemented campus-wide are Virginia Tech's brochure addressing sources of unrecognized bias and its dual-career hiring policy. The brochure addresses sources of unrecognized bias in the hiring process. Presentations using feedback from newly hired faculty and research into hiring bias have been developed and presented to search committees. A respected faculty member, rather than an administrator, makes the presentations, which is key to the success of this program. Search committees, like other faculty groups, are much more receptive to advice and guidance when it is presented by one of their peers rather than by administrators, though the higher administration must also support these efforts if they are to achieve the highest impact.

Because of its rural location, Virginia Tech traditionally has dealt extensively with dual-career hiring, a situation that has a disproportionate impact on women faculty. The importance of this topic came to light as a result of discussions at a university-wide workshop and within the AdvanceVT leadership team and policy work group. The Office of the Provost codified what had been going on informally for years and researched how other universities handle this issue, with both informal and written policies. The provost's office then drafted a policy for Virginia Tech, with input from the faculty senate's Commission on Faculty Affairs, the administration, and department heads. The policy states that the university is committed to dealing with dual-career hiring and will work with candidates to find accommodation, but it does not guarantee positions to faculty partners. Interim funding for the dual hire is available in some cases from the Office of the Provost. The policy was presented to the university's Board of Visitors and has been posted on the provost's office web site. Representatives of AdvanceVT have met with department heads across the university to inform them of the policy. AdvanceVT is working with the personnel office and other resources across the university to provide increased assistance with dual-career hiring and has found that simply making the dual-hire policy information available to each faculty candidate early in the hiring process eases the discussion later on and speeds the resolution of the dual-hire scenario. This policy can reduce the negative impact on partners through a positive resolution supported by an institution-wide policy. Although women have been the most affected in the past, this policy improves the process for all new hires and should be viewed as a means to increase excellence in hiring for all departments.

The Potential Faculty Candidate Outreach, also developed by AdvanceVT, has yielded surprisingly positive outcomes. This program brings women graduate students, postdoctoral research fellows, and junior faculty to campus for seminars outside of the regular faculty search process. Even when there are no specific positions open at the time of these visits, departments want to develop relationships with young women who are potential faculty candidates. AdvanceVT cohosts the visits with individual departments, pays for their travel, introduces them to potential collaborators, hosts a lunch, and arranges for them to meet with graduate students so they can network and talk about their own experiences. Twelve potential candidates have been brought in for visits, two of whom were subsequently hired by Virginia Tech. These hires would not have happened without the intervention of the ADVANCE program. Continued practice of this outreach program will result in an extensive network of young women scientists and junior faculty, not just at Virginia Tech, but also at other institutions as relationships develop between the visitors and the Virginia Tech faculty and students. A similarly proactive program, described elsewhere in this book, has been developed by the University of Puerto Rico–Humacao, which mentors promising undergraduate women through the graduate school application process in the hope that they will eventually return to join the faculty.

Conclusions

This chapter highlights just a few of the low-cost programs implemented by ADVANCE institutions. These programs operate at every level, connecting faculty to faculty, supported by the university administrations, and steadily transforming university communities.

It is likely that there are informal advising relationships between senior and junior faculty at most institutions. By implementing formal advising relationships and workshops, as well as sponsoring venues for informal gatherings, ADVANCE can capitalize upon the experience and wisdom of those who have a natural talent for advising others. These faculty can teach others how to effectively advise and mentor, and reach a wider audience by participating in formal mentoring and workshop activities. More structured activities provide opportunities for staying on-task to tackle and solve difficult issues, while more informal activities provide a relaxed atmosphere. All are examples of institutional community-building that provide opportunities for expertise and knowledge dissemination across campus and venues for networking between faculty groups within and between departments.

By obtaining direct participation at upper administration levels, ADVANCE raises awareness of the confusion and misconceptions surrounding tenure and promotion among the faculty and catalyzes the administration to dispel myths and misinformation, thereby increasing transparency in the promotion process and improving advising. ADVANCE has been able to improve communication between university administrations and faculty in the pipeline, increase the visibility and accessibility of the resources these offices have to offer, and catalyze changes in university policies.

Whether conducted in formal workshop settings or in informal brown-bag activities, programs focused on deans and department chairs have the net effect of transforming entire colleges and departments from the leadership on down, disseminating "best practices" across the campus while building leadership skills and reinforcing leadership communities. For institutional transformation to have the greatest long-term impact, activities directed at academic leaders must be sustained.

Institutional transformation can be further enhanced by addressing instances of bias and less-than-positive language. Although attitude and outlook often seem intractable and subjective, influencing them can result in genuine institutional transformation, especially with top-down administration support. These proactive examples, from language change to hosting potential faculty and building research community networks, cut to the heart of the attitude transformation that ADVANCE strives to promote.

NOTE

Thanks to the following people for their contributions to this chapter: Pam Hunt and Lisa Frehill, New Mexico State University; Joyce Yen and Eve Riskin, University of Washington; Vita Rabinowitz and Virginia Valian, Hunter College, City University of New York; Janet Malley and Danielle LaVaque-Manty, University of Michigan; Barb Silver, University of Rhode Island; Mary Lynn Realff, Georgia Institute of Technology; Peggy Layne, Virginia Polytechnic Institute and State University; Mary Ellen Jackson, University of Maryland, Baltimore County; Patricia Rankin, University of Colorado; Mary Feng, Utah State University. (Affiliations accurate at the time of this writing.)

Written while on a AAAS/NSF Science and Engineering Policy Fellowship at the National Science Foundation, 2005–6.

Appendix

ACES: Academic Careers in Engineering and Science
http://www.case.edu/admin/aces/coaching.htm
This web site provides information on the coaching program described by Bilimoria et al. in chapter 12, along with a coaching template for chairs and a coaching template for women faculty.

ADEPT: Awareness of Decisions in Evaluating Promotion and Tenure
http://www.adept.gatech.edu/
Developed and made available by the Georgia Institute of Technology, this downloadable application, described by Fox et al. in chapter 11, contains case studies and related activities appropriate for group discussion or individual use by candidates, members of committees, and other faculty. An extensive bibliography on bias in evaluation is also available.

ADVANCE Distinguished Lecture Series (ADLS)
http://www.k-state.edu/advance/SeminarsEvents/distinguished_series.htm
This web site provides resources for and reports from the Kansas State University networking program described by Dyer and Montelone in chapter 4.

ADVANCE Portal
http://www.advance-portal.net
Links to research, programs, and announcements contributed by all nineteen ADVANCE institutions are available here, along with links to home pages of all ADVANCE IT awardees.

The Architecture of Inclusion: Advancing Workplace Equity in Higher Education
http://papers.ssrn.com/sol3/papers.cfm?abstract_id=901992
A paper by Susan Sturm about organizational catalysts, related to chapter 16, is available here.

The Center for Research on Learning and Teaching (CRLT) Players
http://www.crlt.umich.edu/theatre/theatre.html
Information about the interactive theater program described by LaVaque-Manty et al. in chapter 13 is available here.

The Committee on the Status of Women in the Economics Profession
http://www.cswep.org
This is the web site for the organization that created the discipline-wide mentoring program described in chapter 10 by Croson and McGoldrick.

Gender Equity Project Workshop Materials
http://www.hunter.cuny.edu/genderequity/workshopmaterials.html
Materials used in the workshops offered by Hunter College's sponsorship program, described by Rabinowitz and Valian in chapter 7, are available on this web site.

New Mexico State University—NSF ADVANCE Indicators
http://www.nmsu.edu/~advprog/Indicators.htm
This page contains links to a toolkit for collecting the NSF indicator data described by Frehill et al. in chapter 18, in both Word and PDF formats.

STRIDE: Strategies and Tactics for Recruiting to Improve Diversity and Excellence
http://sitemaker.umich.edu/advance/stride
This page contains link to PowerPoint presentations, a faculty recruitment handbook, and other materials developed by the recruitment committee described by Stewart et al. in chapter 9.

University of Washington ADVANCE Leadership Workshops
http://www.engr.washington.edu/advance/workshops/
A wide range of materials used in the University of Washington ADVANCE Program's Leadership Workshops, described by Harle in chapter 19, is available here, along with a link to a paper on implementing leadership workshops for department chairs.

University of Washington ADVANCE Transitional Support Program
http://www.engr.washington.edu/advance/tsp.html
Information about the University of Washington's Transitional Support Program, described by Riskin et al. in chapter 8, is available here. The web site includes a sample budget and description of the application process.

Contributors

KELLY ANDRONICOS currently serves as the Executive Administrator for the Center for the Study of Inequality at Cornell University. In this role, Andronicos manages all aspects of an interdisciplinary program that awards the Inequality Concentration Certificate to graduates who have successfully completed a course curriculum oriented toward the study of inequality. In her four years at the University of Texas at El Paso, she worked in the area of faculty development, serving as the Coordinator for the Faculty Mentoring Program for Women, the Center for Effective Teaching and Learning, and the ADVANCE Institutional Transformation grant. Through her close work with faculty, and junior women faculty in particular, Andronicos gained unique insight into the daily challenges and rewards that accompany those who choose a life in the professoriate.

SARA BENÍTEZ is a sociologist and teaches courses on research methods, social theory, and participatory action research at the University of Puerto Rico at Humacao. She is co–principal investigator of the Advance-UPRH NSF program, a researcher for the Prevention and Family Strengthening Alliance Program, and is involved in research projects on systematizing community-based participation. She coordinated the Puerto Rican University women's research network on gender issues, and a UPR research exchange program between UPR faculty and peers in the Caribbean. She has provided leadership to university projects in curriculum development, strategic planning, and self-studies. She was Interim Dean for Academic Affairs, Chair of the Social Science Department, and Director of the Office of Planning and Development. She is a member of the University Senate and of the Administrative

341

Board. She is a feminist and the Director of the first UPR Violence Against Women Prevention Program funded by the Office on Violence Against Women, Department of Justice. She was coeditor of Puerto Rico's Beijing National Report on the Status of Women in Puerto Rico, and is a member of CALCASA, the National Campus Advisory Board, the Sisters of Color Ending Sexual Assault Board, and the Puerto Rico Coalition Against Domestic Violence and Sexual Assault.

DIANA BILIMORIA is Associate Professor of Organizational Behavior at the Department of Organizational Behavior, Weatherhead School of Management, Case Western Reserve University. She received her Ph.D. in Business Administration from the University of Michigan. She is a co-investigator on Case's ADVANCE grant. She served as Editor of the *Journal of Management Education* during 1997–2000. Her research focuses on institutional governance and leadership, gender and diversity in organizations, and management education. She has published articles and book chapters in journals such as the *Academy of Management Journal, Corporate Governance, Group and Organization Management, Human Relations, Group and Organization Management, Journal of Management Education,* and *Women in Management Review,* and in edited volumes such as *Women in Management: Current Research Issues, Women on Corporate Boards of Directors,* and *Advances in Strategic Management.* She is a coeditor of the 2007 *Handbook of Women in Business and Management* (Elgar). She serves as an organizational consultant and management educator for private, public, and nonprofit organizations, as well as an executive coach for individuals. She has received awards for doctoral teaching and professional leadership and service.

SUZANNE G. BRAINARD, Ph.D. is the Executive Director of the Center for Workforce Development at the University of Washington. She is an Affiliate Professor in Technical Communication in the College of Engineering and in the Department of Women Studies in the College of Arts and Sciences. She is one of three cofounders of the Women in Engineering Programs and Advocates Network, and the immediate Past-President. She is past chair of the congressionally mandated Committee on Equal Opportunity in Science & Engineering and served on the National Academy of Engineering Committee on Diversifying the Engineering Workforce and the AAAS [American Association for the Advancement of Science] National Mentoring Committee. She is a Fellow of AAAS and AWIS [Association for Women in Science] and is the recipient of the 2001 Maria Mitchell Women in Science Award. Her research has focused on longitudinal studies examining issues of retention and advancement in engineering and science and the workforce, institutional climate studies at the University of Washington, and national climate surveys in engineering and science. She is a co–principal investigator on the UW ADVANCE grant.

JEAN-LOU A. CHAMEAU is currently President of the California Institute of Technology. At the time this chapter was written, he was Provost and Vice President for Academic Affairs and a Georgia Research Alliance Eminent Scholar at the Georgia Institute of Technology. Previously, he served as dean of the College of Engineering. Dr. Chameau is working to make Georgia Tech a worldwide model for interdisciplinary activities, technology innovation and entrepreneurship, sustainable technology, and a catalyst for economic development. He is placing a strong focus on efforts to improve the educational experience of students, increase diversity on the campus, and foster entrepreneurship and international opportunities for faculty and students. Dr. Chameau received his secondary and undergraduate education in France, and graduate education in civil engineering from Stanford University. He was on the civil engineering faculty at Purdue University from 1980 to 1991, and was director of the School of Civil and Environmental Engineering at the Georgia Institute of Technology from 1991 to 1994. In 1994–95 he was the President of Golder Associates, Inc. He currently serves on the boards of directors for MTS Systems Corporation and Prime Engineering.

CAROL COLATRELLA is Professor of Literature in the School of Literature, Communication, and Culture at Georgia Tech; Co-director of the Georgia Tech Center for the Study of Women, Science, and Technology (WST Center); and Georgia Tech NSF ADVANCE Program Director. She also serves as Executive Director of the Society for Literature, Science, and the Arts and editor of the SLSA newsletter *Decodings*. Her scholarly interests focus on the cultural study of nineteenth- and twentieth-century American and European literary, historical, and scientific narratives. Her publications include *Evolution, Sacrifice, and Narrative: Balzac, Zola, and Faulkner* (1990), *Cohesion and Dissent in America* (coedited with Joseph Alkana, 1994), *Literature and Moral Reform: Melville and the Discipline of Reading* (2002), and articles in journals including *Nineteenth-Century French Studies, Comparative Literature,* and *American Literary History*. She is writing a book analyzing popular culture representations of women engaging with science and technology, *Toys and Tools in Pink: Cultural Narratives of Gender, Science, and Technology*.

RACHEL CROSON is Professor at the University of Texas, Dallas. Her research is in the field of experimental and behavioral economics, and focuses on behavior in strategic (game-theoretic) situations. In particular, her research integrates regularities and findings from psychology into laboratory and field experiments in economics. She has served on the NSF Economics Panel and the NSF ADVANCE panel. She is currently serving her second term on the Board of the Committee on the Status of Women in the Economics Profession, and in 2001 spearheaded a collective NSF ADVANCE Leadership grant application sponsoring the workshops described here. She plans and runs the CeMENT national mentoring workshops.

RUTH A. DYER is Associate Provost and Professor of Electrical and Computer Engineering at Kansas State University. She holds B.S. and M.S. degrees in biochemistry from Kansas State and a Ph.D. in Mechanical Engineering from the University of Kentucky. She joined the faculty at Kansas State in 1983. Her research has been in the areas of digital signal processing, Hadamard transform spectrometry, and biomedical applications of ultrasonics. She served as advisor to the Kansas State chapter of the Society of Women Engineers and initiated the annual Girl Scout Day sponsored by the SWE chapter. She was selected as a Fellow of the American Council on Education in 2003–4 and spent a year at Ohio State University as part of the fellowship. In her current position, she is the provost's Senior Advisor on Gender Issues, and oversees the Office of Planning and Analysis, Assessment, and Summer School. She also is principal investigator on several grants related to human resource development in STEM disciplines.

MARY FRANK FOX is NSF Advance Professor, School of Public Policy, and Co-director of the Center for the Study of Women, Science, and Technology, at Georgia Institute of Technology. She is Co-principal Investigator of Georgia Tech's NSF ADVANCE program. Her research focuses upon gender, science, and academia—and has encompassed analyses of education and educational programs, collaborative practices, salary rewards, publication productivity, social attributions and expectations, and academic careers in science and engineering. Her work appears in over forty scholarly and scientific journals, books, and collections. Her current research projects include a study of programs for women in science and engineering, a longitudinal study of women faculty in computer science, funded by NSF, and the GT ADVANCE research program. She is associate editor of *Sex Roles: A Journal of*

Research; member of the editorial advisory panels of *Social Studies of Science;* and coeditor of the new book series Women, Gender, and Technology, published by University of Illinois Press. She was awarded the SWS Feminist Lecturer 2000 (for a "feminist scholar who has made a commitment to social change"), and the 2002 WEPAN (Women in Engineering Programs) Betty Vetter Research Award (for "notable achievement in research on women in engineering").

LISA M. FREHILL holds a B.S. in industrial engineering from General Motors Institute (now Kettering University) and an M.A and Ph.D. in sociology from the University of Arizona. Since the completion of her studies in 1993, she has been a member of the New Mexico State University Department of Sociology and Anthropology. Her research and teaching focus on research methodology, evaluation, and race/ethnicity, class, and gender as determinants of educational and occupational outcomes. Her evaluation experience includes Camino de Vida Center for HIV Services; New Mexico Alliance for Graduate Education and the Professoriate; Planned Parenthood of New Mexico; and the New Mexico Alliance for Minority Participation "Bridge to the Doctorate" Program. Since 2002, Frehill has been working with many ADVANCE programs, first as the principal investigator and Program Director of NMSU's ADVANCE award (2002–5), then as Program Director of the University of California at Irvine's ADVANCE Program (2005–6). In addition, she is a member of the University of Texas at El Paso and Utah State University's ADVANCE Advisory Boards and has provided technical support related to data collection and analysis to the University of Puerto Rico at Humacao, University of Rhode Island, and the University of Colorado, Boulder ADVANCE programs. In July 2006, she became the Executive Director of the Commission on Professionals in Science and Technology (CPST).

LEE HARLE received her B.S. in physics from Indiana University, Bloomington, in 1996, and her M.S.E. and Ph.D. degrees in electrical engineering from the University of Michigan, Ann Arbor, in 1998 and 2003, respectively. Her thesis focused on micromachined cavity resonator filters in silicon. She was a research fellow at the University of Michigan for one year following the receipt of her Ph.D., conducting research in gallium arsenide packaging of silicon monolithic microwave integrated circuits for communication systems applications. From September 2004 through August 2005, Dr. Harle was an AAAS/NSF Science and Engineering Policy Fellow at the National Science Foundation in the Computer and Information Science and Engineering Directorate. While at NSF, she worked on a variety of funding programs including secure sensing networks, "Science of Design" of software systems, and the ADVANCE program. Currently, Dr. Harle is a Research Engineer at General Dynamics—Advanced Information Systems in Ypsilanti, Michigan, where she works in the area of radar and antenna systems and designs.

MARGARET M. HOPKINS is an Assistant Professor of Management at the University of Toledo teaching courses in leadership and organizational behavior. For the past few years, she has worked with the Executive Education programs and the MBA program at the Weatherhead School of Management, Case Western Reserve University, teaching classes in Leadership Assessment and Development and Organizational Behavior. She has taught Leadership courses in the Masters in Management Program at Ursuline College, and a course in Strategic Planning for Nonprofit and Public Sector Organizations in the graduate program at Cleveland State University. Margaret's areas of research interest include leadership and leadership development, executive coaching, and gender and diversity. She holds a Ph.D. in Organizational Behavior from the Weatherhead School of Management, Case Western Reserve University, a master of science degree in organizational development from CWRU, and a B.S. in psychology from Boston College. She also has an organizational development consulting practice with a specialization in the area of executive coaching. She is actively engaged in the community and recently served as the Chair and the Vice Chair of the Board of Education for the Cleveland Municipal School District from 1998 to 2005.

CECILY JESER-CANNAVALE is the Research Analyst for NSF ADVANCE program at New Mexico State University. She began this position upon completion of her master of arts degree in sociology from New Mexico State University in 2003. She also received her bachelor of arts degree from New Mexico State University in 2001. Research interests include gender equity, social class, and organizational behavior, and she enjoys presenting this research at conferences.

C. GREER JORDAN is a Ph.D. candidate in the Department of Organizational Behavior, Weatherhead School of Management, Case Western Reserve University. She holds an M.B.A. from the University of Michigan and a B.S.E.E. from the University of Detroit. Her research interests include work environment and culture, women leaders, and research methods. She is part of the research staff on Case's ADVANCE grant. She is an adjunct faculty member for the Cleveland State University Diversity Management Program, teaching Assessing, Measuring, and Evaluating Diversity Interventions. She has also been a teaching assistant for an M.B.A.-level Organizational Behavior course at Case. She has served as a reviewer for The *Journal of Action Research*. She provides consulting and facilitation services for organizations focusing on the areas of strategic planning and change management, and serves as Research Director for D. L. Plummer and Associates, a diversity and organizational change management consulting firm. Prior to pursuing her Ph.D., Ms. Jordan held several engineering and management positions at Ford Motor Company and General Motors.

SHEILA EDWARDS LANGE is currently the Associate Director for Research in the Center for Workforce Development at the University of Washington. She manages graduate student mentoring programs and the internal evaluation of the UW ADVANCE program, and was recently appointed to the Women in Engineering Program Advocates Network Board. Her research focuses on underrepresented student access to graduate education, mentoring and professional development, gender in science and engineering, and assessment. She earned a bachelor's degree in social ecology from the University of California at Irvine and a master's degree in public administration from the University of Washington. She is currently a doctoral candidate in Educational Leadership and Policy Studies at the University of Washington. Her dissertation topic is "The Master's Degree as a Critical Transition in Science and Engineering Doctoral Education."

DANIELLE LAVAQUE-MANTY received her Ph.D. in political science from the University of Michigan in 1999. Her dissertation explored the role of ascriptive norms and identity politics in conflicts between indigenous peoples and first-world nations, particularly with respect to anthropological research, the management of museums, and the acquisition and display of indigenous bones and grave goods. She completed an M.F.A. in creative writing at Ohio State University in 2006. She works for NSF ADVANCE at the University of Michigan.

JANET E. MALLEY is currently Associate Director of the Institute for Research on Women and Gender at the University of Michigan and directs the evaluation of the UM ADVANCE project, supported by the

NSF ADVANCE program on Institutional Transformation. She received her Ph.D. in personality psychology from Boston University and completed a postdoctoral fellowship at the University of Michigan's Institute for Survey Research. Her research interests are in the area of adult development with a special focus on women's lives, looking especially at how the process of development may be mediated by individual life experiences as well as more broadly based work and family roles.

DAVID MCDOWELL is Regents' Professor at Georgia Institute of Technology. Holding the Carter N. Paden, Jr. Distinguished Chair in Metals Processing, McDowell's primary appointment is in the George W. Woodruff School of Mechanical Engineering, with a joint appointment in the School of Materials Science and Engineering. He serves as Director of the Mechanical Properties Research Laboratory and as Chair of the Georgia Tech Materials Council. McDowell's research focuses on combined experimental and computational methods to develop physically based constitutive models for nonlinear and time-dependent behavior of materials, with emphasis on wrought and cast metals. His interests include multiscale modeling of material behavior in support of multifunctional materials design. Serving as Chair of the Promotion and Tenure ADVANCE Committee at Georgia Tech from 2002 to 2003, McDowell led the effort to canvass promotion and tenure practices within various academic units, survey faculty perceptions related to career advancement, and develop best-practices guidelines and web-based tools for faculty evaluation committees to explore potential sources of bias in promotion and tenure decisions.

KIMMARIE MCGOLDRICK is Professor of Economics in the Robins School of Business at the University of Richmond in Virginia. She has taught Principles, Statistics, Labor, Industrial Organization and Public Policy, Economics of Poverty and Discrimination, Women and Gender Issues in Economics, and Introduction to Feminist Theory. She is passionate about developing research skills in her students and has developed and is currently teaching the economics department senior capstone experience in which students use the theoretical and applied economic models that they have been exposed to in their economics courses. Over the past six years she has coordinated a number of workshops including Service Learning in Economics, the Annual Teaching Workshop held in Wilmington, NC, and CeMENT regional mentoring workshops. She has joined the staff of the Teaching Innovations Program, an NSF-funded project designed to improve undergraduate education in economics by offering instructors an opportunity to expand their teaching skills and participate in the scholarship of teaching and learning. Her research interests include pedagogical practices such as service learning and cooperative learning and more traditional empirical research measuring market information using frontier analysis as applied to unique subjects such as efficiency of play in the Women's National Basketball Association versus the National Basketball Association.

BETH A. MONTELONE is Associate Dean of the College of Arts and Sciences and Professor of Biology at Kansas State University. She earned a B.S. in biology from Rensselaer Polytechnic Institute and a Ph.D. in biology from the University of Rochester. She completed postdoctoral training at the University of Miami School of Medicine and the University of Iowa prior to joining the faculty at Kansas State in 1988. Her research has been in the area of the molecular genetics of DNA repair and mutagenesis. She has also been active in curriculum reform, K–12 teacher outreach, and in promoting undergraduate research opportunities. She was a founding member and past president of the Kansas Flint Hills chapter of the Association for Women in Science. In her current position, she is responsible for gender equity programs and facilitating research activities within the college and is principal investigator on several grants related to human resource development in STEM disciplines.

JOYCE MCCARL NIELSEN is Professor of Sociology and Associate Dean for Social Sciences, College of Arts and Sciences, University of Colorado at Boulder. In earlier work she helped define the field of gender studies with her texts *Sex and Gender in Society: Perspectives on Stratification* (1978, Wadsworth Publishing; 2nd ed. 1990, Waveland Press) and *Feminist Research Methods: Exemplary Readings in the Social Sciences* (1990, Westview Press). She is a co-principal investigator and active researcher on the University of Colorado's NSF ADVANCE LEAP (Leadership Education for Advancement and Promotion) grant.

DEBORAH A. O'NEIL is a Senior Lecturer on the faculty of the Organizational Behavior Department at the Weatherhead School of Management, Case Western Reserve University. She teaches classes in leadership development and organizational behavior in the M.B.A., Executive MBA and Executive Education programs. Dr. O'Neil has published articles on career development, the use of coaching behaviors in management education, and the importance of emotional intelligence in developing leadership skills for life. Her research is focused on career-in-life development, life transitions, women leaders, and the positive impact of coaching and mentoring relationships. In June 2005 she was awarded a National Science Foundation ACES (Academic Careers in Sciences and Engineering) Opportunity Grant to present her research at the European Group and Organization Studies conference in Berlin, Germany. Dr. O'Neil holds a doctorate in organizational behavior from the Weatherhead School of Management at Case Western Reserve University, a master of science degree in organizational development from American University in Washington, DC, and bachelor's degrees in secondary education, journalism and English from the University of Rhode Island.

SUSAN R. PERRY is a Senior Research Associate for the Academic Careers in Engineering and Science program, and is affiliated with the Organizational Behavior Department at the Weatherhead School of Management, Case Western Reserve University. Dr. Perry has published articles on job satisfaction and career success, and has also conducted research in the areas of personality and work drive. Her current research is focused on gender and academic careers, and the impact of individual, relational, and organizational structures on career success. Dr. Perry holds a doctorate in psychology from the University of Tennessee, a master of arts degree in psychology from Kent State University, and a bachelor's degree in psychology from King College.

EVELYN POSEY is the Dorrance D. Roderick Endowed Professor and Chair of the University of Texas at El Paso Department of English and principal investigator on the NSF ADVANCE Institutional Transformation for Faculty Diversity initiative. A specialist in rhetoric and writing studies, she has served as Director of English Education; Director of the West Texas Writing Project, a site of the National Writing Project; Associate Dean of Liberal Arts; and Associate Vice President for Academic Affairs. Posey has published in *Guide to Writing Center Theory and Practice, Computers and Composition,* the *Journal of Developmental Education, Teaching English in the Two-Year College,* and the *Writing Center Journal.* With Kate Mangelsdorf, she has coauthored two composition textbooks: *Choices: A Basic Writing Guide with Readings,* 3rd ed. (2003) and *Discoveries: A Step-by-Step Guide to Writing Paragraphs and Essays* (2006), both published by Bedford/St. Martin's Press. For her work in the El Paso community, she was awarded the YWCA REACH Award for Outstanding Achievement in the Business Community (2000), and was inducted into the El Paso Women's Hall of Fame (2001) and the El Paso Technology Hall of Fame (2002).

KATE QUINN is a doctoral candidate in the Graduate Program in Higher Education at the University of Washington. Her research interests include higher education policy, climate, and culture as they structure work-life balance for faculty and students, as well as the professional socialization and leadership development of graduate students and faculty. She received her M.Ed. in educational leadership and policy studies from the University of Washington and her B.A. in English from the State University of New York at Geneseo. She is a graduate research assistant for UW ADVANCE Center for Institutional Change.

VITA C. RABINOWITZ is Acting Provost and Professor of Psychology at Hunter College of the City University of New York, where she has taught for twenty-seven years. She is also a member of the Social-Personality Subprogram of the Psychology Doctoral Program of the City University of New York. She received a B.A. in psychology and English from Douglass College of Rutgers University, and her masters and doctorate in social psychology from Northwestern University. Since 2002, Professor Rabinowitz has codirected Hunter's Gender Equity Project. She is currently a member of the New York Academy of Sciences' Women Investigators Network. Professor Rabinowitz has written extensively in the areas of gender and achievement, gender bias in science, particularly health research, and research methodology. Her coauthored textbook on the psychology of women, *Engendering Psychology: Women and Gender Revisited,* published by Allyn and Bacon, has just appeared in its second edition.

IDALIA RAMOS is an Associate Professor in the Department of Physics and Electronics of the University of Puerto Rico, Humacao Campus. She is the principal investigator for the ADVANCE Program at UPR-Humacao. Her current research interests include the synthesis and characterization of nanostructures and their use as materials for sensors and actuators and strategies to increase the participation of Puerto Rican women in science. Ramos is also the principal investigator of an NSF-sponsored Partnership for Research and Education in Materials Science, a collaborative effort between the University of Puerto Rico and the University of Pennsylvania. As a faculty member at Humacao she initiated and implemented efforts to encourage women students to pursue careers in physics and other sciences, including workshops for high school and undergraduate women, conferences, providing undergraduate research opportunities, and mentoring. She has also served as co–principal investigator for a Program to Stop Violence Against Women on campus and a member of the UPR-Humacao Women's Affairs Committee.

PATRICIA RANKIN is Professor of Physics and Associate Dean for the Natural Sciences at the University of Colorado, Boulder. She was born in Liverpool, England, and went to an all-girls school from age eleven to age eighteen. She believes that this early socialization has allowed her to enjoy an unusual career and follow her interests both in particle physics research and in broader societal issues such as the failure of the scientific workforce to diversify. She did both her undergraduate and her graduate work at Imperial College, London, and then moved to the United States to become a postdoctoral researcher at the Stanford Linear Accelerator Center. She is interested in the physics of heavy quarks and the symmetries of nature. After getting tenure she worked for two years in Washington, DC, as a program officer at the National Science Foundation. Returning to Boulder, she took on a leadership role on campus in working to address the lack of representation of women in STEM fields and in promoting best practices for departmental and campus leaders

MARY LYNN REALFF, Associate Professor of Polymer, Textile, and Fiber Engineering at Georgia Institute of Technology, is currently Program Officer for the Materials Processing and Manufacturing Program of the National Science Foundation. At Georgia Tech, she taught graduate and undergraduate courses in the mechanics of textile structures and polymer science areas. Dr. Realff has made a significant contribution to the understanding of the mechanical behavior of woven fabrics, and currently conducts research on the analysis and design of high performance fibers. Dr. Realff is Co-Director of the Center for the Study of Women, Science and Technology, and facilitates a student/industry mentoring program which matches industry mentors with Georgia Tech undergraduate students for mentoring relationships. From 2002–2005 Dr. Realff was Director of the Georgia Tech NSF ADVANCE grant. Dr. Realff received her B.S. from Georgia Tech and Ph.D. from MIT in Mechanical Engineering and Polymer Science and Technology. She is the mother of Alexander (age 10) and Penelope (age 8) and is married to Dr. Matthew Realff.

CHRISTINE REIMERS, Ph.D., has taught humanities, writing and rhetoric, world literature, French, and philosophy in a variety of settings, from large American Research I universities, to small colleges, to universities in France and Japan. Her twenty years of classroom experience teaching students from diverse backgrounds, combined with eleven years of faculty development experience at the University of North Carolina–Chapel Hill, Indiana University–Bloomington, and currently the University of Texas at El Paso have provided the opportunity to deepen her understanding and communication of the strategies best suited for effective learning in higher education. She works with faculty and graduate students from all disciplines to enhance course design that emphasizes discovery learning and student inquiry. Her workshops are designed for active engagement of the participants, modeling and putting into practice the ideas she is teaching, whether they be critical thinking, using teamwork in the classroom, lecturing effectively, or planning for active learning. Dr. Reimers also provides more comprehensive professional development of faculty in order to better prepare and sustain them for their careers in academe. She offers programs with a focus on new faculty, and those supporting the continuing professional development for later "stages" in a faculty life.

EVE A. RISKIN received the B.S. degree in electrical engineering from MIT and M.S. degrees in electrical engineering and operations research and a Ph.D. degree in electrical engineering, all from Stanford University. Since September 1990, she has been at the University of Washington, where she is now Associate Dean of Organizational Infrastructure, Professor of Electrical Engineering, and Director of the ADVANCE Center for Institutional Change. Her research interests include image compression and image processing. She was awarded the National Science Foundation Young Investigator Award in 1992 and the Sloan Research Fellowship in 1994.

SUE V. ROSSER received her Ph.D. in zoology from the University of Wisconsin–Madison in 1973. Since July 1999, she has served as Dean of Ivan Allen College, the liberal arts college at Georgia Institute of Technology, where she is also Professor of History, Technology, and Society and Professor of Public Policy, and holds the Ivan Allen Dean's Chair in Liberal Arts and Technology. She has edited collections and written approximately 115 journal articles on the theoretical and applied problems of women and science and women's health. Author of nine books, her latest is *The Science Glass Ceiling: Academic Women Scientists and their*

Struggle to Succeed (2004) from Routledge. She has held several grants from the National Science Foundation, and currently serves as co–principal investigator on Georgia Tech's ADVANCE grant. She is an AWIS · [Association for Women in Science] Fellow, AAAS [American Association for the Advancement of Science] Fellow, 2003 recipient of the Betty Vetter WEPAN Research Award, and serves as a 2005–7 Sigma Xi Distinguished Lecturer.

DAWN M. STANLEY is a graduate student in sociology and a research associate for the LEAP project at the University of Colorado at Boulder. She completed her B.A. in 1997 at St. Lawrence University, majoring in theater and sociology. Her dissertation examines the development of the diagnosis of and treatments for post-traumatic stress disorder. In addition to her research on gender and academia, she has also worked with a Colorado-based support group for families of homicide victims.

JEFFREY STEIGER is the Director of the CRLT Theatre Program at the University of Michigan and has been working in theatre and with interactive theatre techniques for over fifteen years. His career as an actor and director centers on the idea of theatre as a compelling agent for social change. Jeffrey has been at the Center for Research on Learning and Teaching since 2000. As Director of the CRLT Theatre Program he creates original scripts, recruits and develops actors, consults with faculty and graduate student instructors on voice and communication issues, and works with academic units to apply theatre to their faculty development needs. Under his direction, the CRLT Theatre Program has become a national resource, performing at campuses and conferences around the country. In the summer of 2005, he designed and led a three-day Institute on Using Interactive Theatre to Improve Institutional Climate. This collaboration between CRLT and the NSF ADVANCE Project at UM brought together thirty-five faculty and faculty developers from eighteen institutions. A similar program was held in June 2006. In addition to his work with the CRLT Players, he writes, directs, and produces original works for the stage. He has also taught courses on performance art, comedy, and acting for nonactors both at UM and in the Ann Arbor public schools. Jeffrey is currently developing scripts on mentoring, international faculty, sexual orientation, race, and other multicultural topics.

ABIGAIL J. STEWART is Sandra Schwarz Tangri Professor of Psychology and Women's Studies at the University of Michigan and director of the

UM ADVANCE project, supported by the NSF ADVANCE program on Institutional Transformation. She is former director of the Women's Studies Program (1989–95), of the Institute for Research on Women and Gender (1995–2002), and former Associate Dean for Academic Affairs in the College of Literature Science and the Arts at the University of Michigan (2002–4). She has published many scholarly articles and several books, focusing on the psychology of women's lives, personality, and adaptation to personal and social changes. Her current research, which combines qualitative and quantitative methods, includes comparative analyses of longitudinal studies of educated women's lives and personalities; a collaborative study of race, gender, and generation in the graduates of a Midwest high school; and research and interventions on gender and science and technology with middle-school-age girls, undergraduate students, and faculty.

SUSAN STURM is the George M. Jaffin Professor of Law and Social Responsibility at Columbia Law School, where her principal areas of teaching and research include employment discrimination, workplace regulation, race and gender, public law remedies, and civil procedure. Her current work focuses on rethinking employment discrimination regulation, addressing complex forms of bias, and examining sites for successful multiracial problem solving. She is a founding member of Columbia University's Presidential Advisory Committee on Diversity Initiatives. Her recent publications include *Who's Qualified? The Future of Affirmative Action* (with Lani Guinier) (Beacon Press, 2001); *Law's Role in Addressing Complex Discrimination* (2005); *Equality and the Forms of Justice* (2004); "Learning from Conflict: Reflections on Teaching About Race and Gender" (*Journal of Legal Education,* 2003); "Lawyers and the Practice of Workplace Equity" (*Wisconsin Law Review,* 2002); *Second Generation Employment Discrimination: A Structural Approach* (Columbia University Press, 2001); "Equality and Inequality" (*International Encyclopedia of Social Sciences,* 2001); and *Race, Gender and the Law in the Twenty-First Century Workplace* (1998). She has a forthcoming article entitled "From Affirmative Action to Institutional Transformation: The Power of Mobilizing Knowledge." She also has developed a web site with Lani Guinier, www.racetalks.org, on building multiracial learning communities.

VIRGINIA VALIAN is Distinguished Professor of Psychology and Linguistics at Hunter College and the Graduate Center of the City University of New York (CUNY). She is a cognitive scientist whose research ranges from first- and second-language acquisition to gender equity.

The National Science Foundation (NSF) currently funds Valian's research in language acquisition and her work on gender. She is codirector of Hunter College's Gender Equity Project, the initiatives of which are described at www.hunter.cuny.edu/genderequity. She has also created web-based tutorials on the role of gender in professional life. The tutorials take the form of slides with voice-over narration and are intended for students, faculty, and administrators everywhere in the world. The tutorials can be accessed at www.hunter.cuny.edu/gender-tutorial. Valian's book *Why So Slow? The Advancement of Women,* published by MIT Press, has been described by reviewers as "compelling," "scholarly and convincing," "accessible and lively," and "a breakthrough in the discourse on gender." She lectures and gives workshops to faculty and professional groups in the United States and abroad.

JOYCE W. YEN received her M.S. and Ph.D. in industrial and operations engineering from the University of Michigan. She received her B.S. in mathematics from the University of Nebraska–Lincoln. Her research interests include decision making and resource allocation under uncertainty (stochastic programming), faculty and graduate student professional development, and women in science and engineering issues. She was awarded the University of Nebraska–Lincoln's 2004 Outstanding Young Alumni Award. Dr. Yen was previously an assistant professor in Industrial Engineering at the University of Washington, Seattle and is currently the Program/Research Manager for the University of Washington's NSF-funded ADVANCE Center for Institutional Change.

Index